SMP Interact

Teacher's guide to Book 7S

CAMBRIDGE UNIVERSITY PRESS

PUBLISHED BY THE PRESS SYNDICATE OF THE UNIVERSITY OF CAMBRIDGE
The Pitt Building, Trumpington Street, Cambridge, United Kingdom

CAMBRIDGE UNIVERSITY PRESS
The Edinburgh Building, Cambridge CB2 2RU, UK
40 West 20th Street, New York, NY 10011–4211, USA
477 Williamstown Road, Port Melbourne, VIC 3207, Australia
Ruiz de Alarcón 13, 28014 Madrid, Spain
Dock House, The Waterfront, Cape Town 8001, South Africa

http://www.cambridge.org

© The School Mathematics Project 2003

First published 2003
Reprinted 2003

Printed in the United Kingdom at the University Press, Cambridge

Typeface Minion   *System* QuarkXPress®

A catalogue record for this book is available from the British Library

ISBN 0 521 53794 0 paperback

Typesetting and technical illustrations by The School Mathematics Project and Jeff Edwards
Photographs by Graham Portlock
Cover image Getty Images/Randy Allbritton
Cover design by Angela Ashton

NOTICE TO TEACHERS
It is illegal to reproduce any part of this work in material form (including photocopying
and electronic storage) except under the following circumstances:
(i) where you are abiding by a licence granted to your school or institution by the
Copyright Licensing Agency;
(ii) where no such licence exists, or where you wish to exceed the terms of a licence, and
you have gained the written permission of Cambridge University Press;
(iii) where you are allowed to reproduce without permission under the provisions
of Chapter 3 of the Copyright, Designs and Patents Act 1988.

Contents

Introduction *5*
Mental methods and starters *7*
1 First bites *20*
2 Symmetry 1 *34*
3 Whole number calculation *40*
4 Oral questions: time *50*
5 Growing patterns *51*
6 Test it! *58*
7 Angle *61*
8 Time *66*
9 Place value and rounding *71*
Review 1 *78*
Mixed questions 1 *78*
10 Using fractions *79*
11 Balancing *83*
12 Health club *86*
13 Understanding decimals *94*
14 Temperature *101*
15 Oral questions: money 1 *105*
16 Coordinates *106*
17 Number patterns *109*
18 Brackets *113*
Review 2 *118*
Mixed questions 2 *118*
19 Rounding decimals *120*
20 Gravestones *124*
21 Oral questions: recipes *128*
22 Work to rule *129*
23 Triangles *135*
24 Decimal calculation *140*
25 Equivalent fractions and ratio *145*
26 Area and perimeter *148*
Review 3 *152*
Mixed questions 3 *152*

27 Spot the rule *153*
28 Chance *158*
29 Oral questions: calendar *162*
30 Number grids *163*
31 Fractions and decimals *174*
32 Oral questions: measures *180*
33 Inputs and outputs *182*
34 Action and result puzzles *189*
35 Perpendicular and parallel lines *191*
Review 4 *195*
Mixed questions 4 *196*
36 Practical problems *197*
37 Comparisons *199*
38 Multiplying and dividing *204*
39 Think of a number *210*
40 Symmetry 2 *216*
41 Hot and cold *224*
42 Investigations *225*
43 Decimals with a calculator *235*
44 Three dimensions *239*
45 Percentage *242*
Review 5 *247*
Mixed questions 5 *247*
46 Fair to all? *249*
47 Transformations *255*
48 Negative numbers *259*
49 Oral questions: money 2 *262*
50 Big numbers *264*
51 Functions and graphs *265*
52 Multiples and factors *269*
53 Know your calculator *274*
54 Chocolate *282*
Review 6 *285*
Mixed questions 6 *285*

The following people contributed to the writing of the SMP Interact key stage 3 materials.

Ben Alldred	Ian Edney	John Ling	Susan Shilton
Juliette Baldwin	Steve Feller	Carole Martin	Caroline Starkey
Simon Baxter	Rose Flower	Peter Moody	Liz Stewart
Gill Beeney	John Gardiner	Lorna Mulhern	Pam Turner
Roger Beeney	Bob Hartman	Mary Pardoe	Biff Vernon
Roger Bentote	Spencer Instone	Peter Ransom	Jo Waddingham
Sue Briggs	Liz Jackson	Paul Scruton	Nigel Webb
David Cassell	Pamela Leon	Richard Sharpe	Heather West

Others, too numerous to mention individually, gave valuable advice, particularly by commenting on and trialling draft materials.

Editorial team
David Cassell
Spencer Instone
John Ling
Paul Scruton
Susan Shilton
Caroline Starkey
Heather West

Project administrator
Ann White

Design
Pamela Alford
Melanie Bull
Nicky Lake
Tiffany Passmore
Martin Smith

Project support
Carol Cole
Pam Keetch
Jane Seaton
Cathy Syred

Special thanks go to Colin Goldsmith.

Introduction

Teaching approaches

SMP Interact sets out to help teachers use a variety of teaching approaches in order to stimulate pupils and foster their understanding and enjoyment of mathematics.

A central place is given to discussion and other interactive work. In this respect and others the material supports the methodology of the *Framework for teaching mathematics*. Questions that promote effective discussion and activities well suited to group work occur throughout the material.

Some activities, mostly where a new idea or technique is introduced, are described only in the teacher's guide. (These are indicated in the pupils' book by a solid marginal strip – see below.)

Materials

There are three series in key stage 3: *Books 7T–9T* cover up to national curriculum level 5; *7S–9S* go up to level 6; *7C–9C* go up to level 7, though schools have successfully prepared pupils for level 8 with them, drawing lightly on extra topics from early in the *SMP Interact* GCSE course.

The year 7 books share much common material – a benefit where mixed attainment groups or broad setting are used for an initial settling-in period, or where the school covers topics in parallel to ease transfer between sets. To help you with your planning, links to common and related material – between *Book 7T* and *7S*, and between *7S* and *7C* – are shown to the right of unit headings (in both the pupils' book and the teacher's guide); for example, unit 2 of *Book 7S* has the links '7T/5, 7C/2', meaning there is common or related material in unit 5 of the less demanding *Book 7T* and in unit 2 of the more demanding *Book 7C*.

All three year 7 books start with a collection of activities called 'First bites', designed to help you get to know pupils and to give them an enjoyable and confident start in mathematics at secondary school.

Pupils' books

Each unit of work begins with a statement of learning objectives and most units end with questions for self-assessment.

Teacher-led activities that are described in the teacher's guide are denoted by a solid marginal strip in both the pupils' book and the teacher's guide.

Some other activities that are expected to need teacher support are marked by a broken strip.

Where the writers have particular classroom organisation in mind (for example working in pairs or groups), this is stated in the pupils' book.

Resource sheets

Resource sheets, some essential and some optional, are linked to some activities in the books.

Practice booklets

For each book there is a practice booklet containing further questions unit by unit. These booklets are particularly suitable for homework.

Teacher's guides

For each unit, there is usually an overview, details of any essential or optional equipment, including resource sheets, and the practice booklet page references, followed by guidance that includes detailed descriptions of teacher-led activities, advice on difficult ideas and comments from teachers who trialled the material.

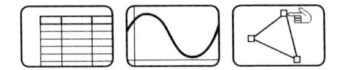

There is scope to use computers and graphic calculators throughout the material. These symbols mark specific opportunities to use a spreadsheet, graph plotter and dynamic geometry software respectively.

Answers to questions in the pupils' book and the practice booklet follow the guidance. For reasons of economy answers to resource sheets that pupils write on are not always given in the teacher's guide; they can of course be written on a spare copy of the sheet.

Assessment

Unit by unit assessment tests are available both as hard copy and as editable files on CD (details are at www.smpmaths.org.uk). The practice booklets are also suitable as an assessment resource.

Mental methods and starters

Mental methods

At various points in the book there are references to, and examples of, mental calculation. Although a longer session focusing on specific skills can occasionally be useful, work on developing mental methods is generally best spread out on a 'little and often' basis. The oral and mental 'starter' (see pages 13–19) is an ideal place for this.

The following suggestions are intended to help you support pupils in developing and extending their methods of mental calculation. All will benefit from seeing and using a variety of methods. But in the end it is accuracy and efficiency which count, so pupils should not be forced to learn a particular method when they have others that work well for them.

Pick up on pupils' own ideas and methods and discuss and compare them. Often the method will vary with the numbers involved. For example, 25 + 19 might be done as 25 + 20 − 1, whereas 25 + 26 might be done as 25 + 20 + 6, or even 2 × 25 + 1.

Vary the language you use. For example, in the case of addition you can use 'add', 'total', 'add to', 'add together', 'increase by', '… more than', etc.

Adding

Unit 3 section A provides a suitable starting point for mental addition and subtraction. The different methods described below can then be looked at in short sessions spaced out over time.

An **unmarked number line** is a very useful model. To start with you could mark just the hundreds or tens (or units or decimals). Pages 20 and 21 of the pupils' book show some number lines marked in tens.

Adding by partitioning into 100s, 10s … (no 'carry')

Examples   45 + 13

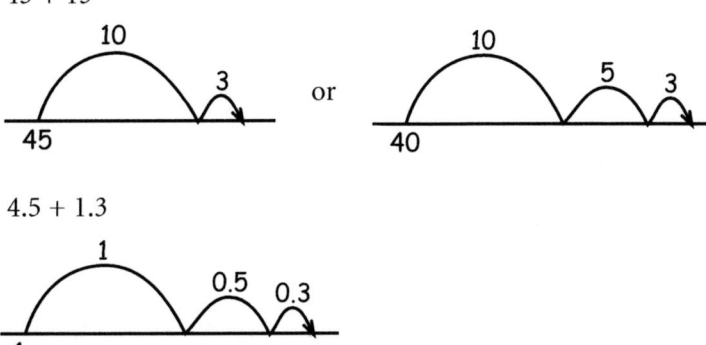

or

4.5 + 1.3

Other examples   83 + 12   146 + 42   161 + 238   7.4 + 2.3   1.24 + 0.31

Adding 9s (or 0.9s) by adding 10s (or 1s) then adjusting

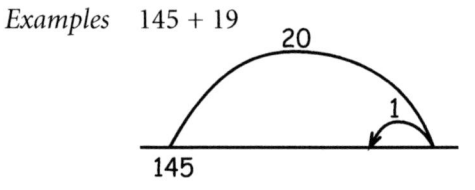

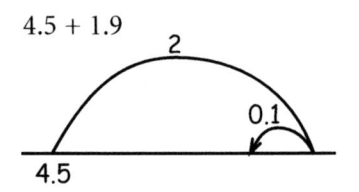

Examples 145 + 19   4.5 + 1.9

£6.27 + £3.99

Other examples   52 + 19   223 + 39   29 + 144   5.6 + 2.9   £16.40 + £5.99

Adding the nearest multiple of 100, 10 … then adjusting

Examples 267 + 96

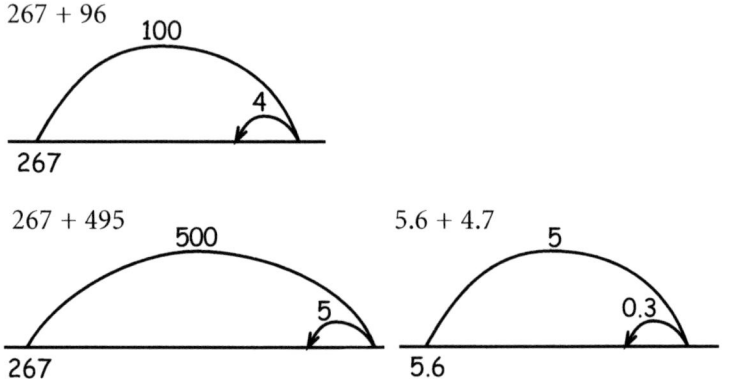

267 + 495   5.6 + 4.7

Other examples   23 + 18   137 + 48   12.5 + 3.7

Adding by partitioning into 100s, 10s … (with 'carry')

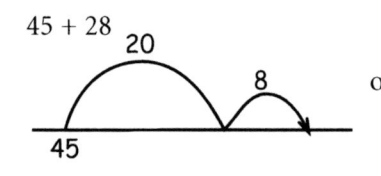

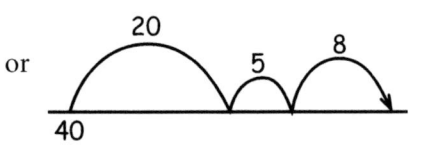

Example 45 + 28   or

Other examples   39 + 12   47 + 24   36 + 56   37 + 25   55 + 17   38 + 32

'Counting on' to a multiple of 100, 10 …

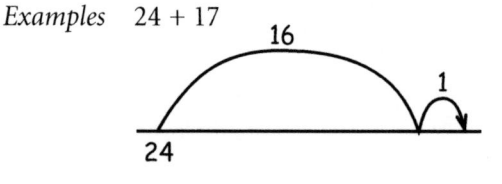

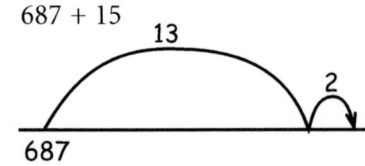

Examples 24 + 17   687 + 15

8 • *Mental methods and starters*

$$8.9 + 0.6$$

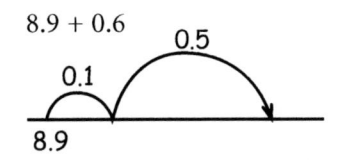

Other examples 139 + 17   247 + 25   3.6 + 0.7

Looking for pairs that make multiples of 100, 10, 1

Examples

21 + 7 + 39
= 60 + 7
= 67

1.2 + 0.3 + 3.8 + 0.4
= 5 + 0.7
= 5.7

Other examples 142 + 29 + 58   4.5 + 1.6 + 7.5

Subtracting

As with addition, an **unmarked number line** is a very useful model.

Counting on

Examples 104 − 98   18.2 − 17.9

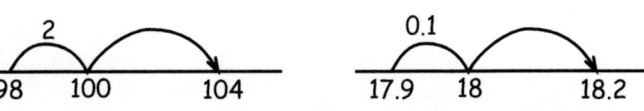

8000 − 2785

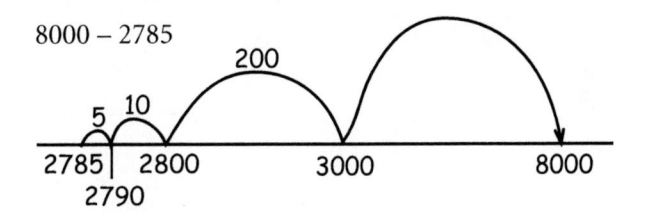

Other examples 35 − 32   41 − 39   172 − 168   590 − 587   5.1 − 4.8
300 − 264   7003 − 6899   8006 − 2993

Subtracting to a multiple of 100, 10 … first

Examples 22 − 7   12.5 − 0.7

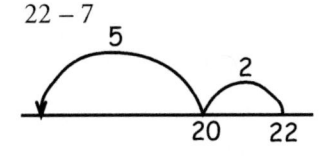

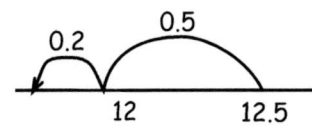

Other examples 43 − 8   161 − 6   324 − 26   9.1 − 0.3

Mental methods and starters • 9

Subtracting 100s, 10s, ... not crossing a multiple of 100, 10, ...

Examples   247 − 16   9.8 − 4.3

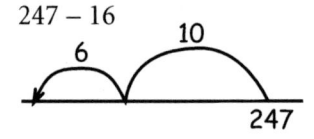

Other examples   474 − 143   89 − 45   7.8 − 2.3   19.9 − 6.2

Subtracting 9s (or 0.9s) by subtracting 10s (or 1s) then adjusting

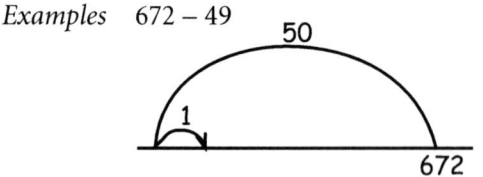

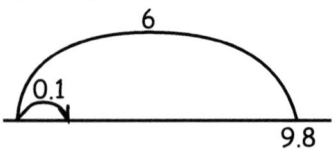

Examples   672 − 49   9.8 − 5.9

£20 − £6.99

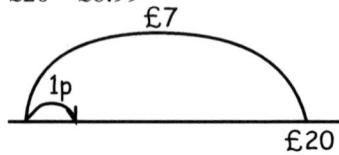

Other examples   56 − 9   178 − 29   5.3 − 1.9   706 − 399

Subtracting the nearest multiple of 100, 10 ... then adjusting

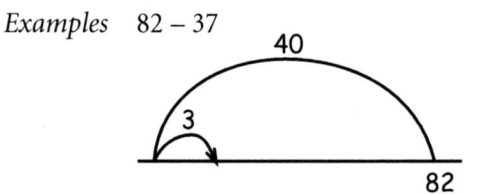

Examples   82 − 37   £19.08 − £5.97

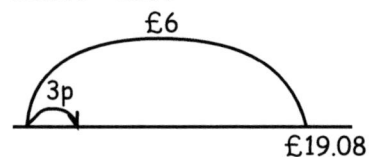

Other examples   173 − 37   4.1 − 2.7   465 − 238   703 − 296   4005 − 1997

Subtracting 100s, 10s, ... crossing a multiple of 100, 10, ...

Examples   162 − 23   5.4 − 2.7

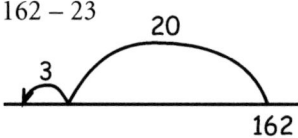

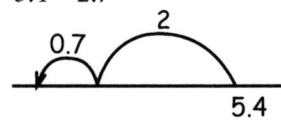

Other examples   73 − 34   141 − 27   465 − 238   7.2 − 3.3

Multiplying

Unit 3 section C provides a suitable starting point.

Doubling by partitioning into 100s, 10s, 1s, …

Examples double 176 = double 100 + double 70 + double 6
= 200 + 140 + 12 = 352

6.8×2 = double 6 + double 0.8 = 12 + 1.6 = 13.6

Multiplying by 4 and by 8 by repeated doubling

To multiply by 4, double and then double again

Example $26 \times 4 = 26 \times 2 \times 2 = 52 \times 2 = 104$

To multiply by 8, double, double and then double again

Example $26 \times 8 = 26 \times 2 \times 2 \times 2 = 52 \times 2 \times 2 = 104 \times 2 = 208$

Multiplying by 6 by multiplying by 2 and 3 (in either order)

Examples $26 \times 6 = 26 \times 2 \times 3 = 52 \times 3 = 156$

$43 \times 6 = 43 \times 3 \times 2 = 129 \times 2 = 258$

$4.5 \times 6 = 4.5 \times 2 \times 3 = 9 \times 3 = 27$

Multiplying by 9 by multiplying by 10 and adjusting

Example 14 times 9 = 14 times 10 then subtract 14 = 140 − 14 = 126

Doubling a number ending in 5 and halving the other number

Examples 16×5 = half of 16 × double 5 = 8 × 10 = 80

$35 \times 14 = 70 \times 7 = 490$

Using factors

Examples $16 \times 51 = 2 \times 8 \times 51 = 2 \times 408 = 816$

$12 \times 35 = 12 \times 5 \times 7 = 60 \times 7 = 420$

$3.2 \times 30 = 3.2 \times 10 \times 3 = 32 \times 3 = 96$

Multiplying by 50 by multiplying by 100 and then halving

Examples 36×50 = half of 36 times 100 = 3600 ÷ 2 (or 18 × 100)
= 1800

$1.38 \times 50 = 1.38 \times 100 \div 2 = 138 \div 2 = 69$

Multiplying by 25 by multiplying by 100 and then dividing by 4

Examples $38 \times 25 = 38 \times 100 \div 4 = 3800 \div 4 = 950$

$3.6 \times 25 = 3.6 \times 100 \div 4 = 360 \div 4 = 90$

Multiplying by numbers 1 more or less than a multiple of 10 or 100

Multiply by the nearest multiple of 10, 100 and then adjust

Examples   13 times 21 = 13 times 20 then add 13 = 260 + 13 = 273
$13 \times 49 = (13 \times 50) - 13 = 650 - 13 = 637$
$13 \times 199 = (13 \times 200) - 13 = 2600 - 13 = 2587$

Multiplying by partitioning into 100s, 10s, …

Examples   86 times 7 = (80 times 7) + (6 times 7) = 560 + 42 = 602
$8.6 \times 7 = (8 \times 7) + (0.6 \times 7) = 56 + 4.2 = 60.2$

Dividing

Unit 3 section D provides a suitable starting point.

Halving by partitioning into 100s, 10s, 1s, …

Examples   half of 158 = half of 100 + half of 50 + half of 8
= 50 + 25 + 4 = 79

half of 5.6 = half of 5 + half of 0.6 = 2.5 + 0.3 = 2.8

Dividing by 4 or 8 by repeated halving

To divide by 4, halve and then halve again

Example $96 \div 4 = 96 \div 2 \div 2 = 48 \div 2 = 24$

To divide by 8, halve, halve and then halve again

Example $96 \div 8 = 96 \div 2 \div 2 \div 2 = 48 \div 2 \div 2 = 24 \div 2 = 12$

Dividing by 6 by dividing by 2 and 3 in either order

Example $96 \div 6 = 96 \div 3 \div 2 = 32 \div 2 = 16$

Using factors

Examples $96 \div 12 = 96 \div 3 \div 4 = 32 \div 4 = 8$
$70 \div 14 = 70 \div 7 \div 2 = 10 \div 2 = 5$

Oral and mental starters

An oral or mental starter can be used for a number of purposes.

- It can **introduce the main topic**, and many of the teacher-led activities described in this guide can be used in this way.
- It can also be an effective way of **revising skills that are needed for the main topic** and can prevent the subsequent lesson 'sagging' when those skills falter. For example, 'coordinate bingo' with negative coordinates (see the guidance for unit 16 section C) could precede the main topic of linear graphs with negative coordinates (unit 51 section D).
- Alternatively a starter can be used to **revise skills that are unrelated to the main lesson**. Some questions and activities in the pupils' book can be adapted for later use as starters. For example, the 'Three digits' activity on page 32 of the pupils' book could be used with any three digits at the beginning of any lesson.

Starter formats

The formats described below have been found very effective and can be adapted to different topics.

Small whiteboards ('show me boards') and markers are invaluable for many types of starter, for example *True or false?*, *What am I?* and *Odd one out*. Pupils write their responses on their board and hold it up, giving you instant feedback on the whole class.

True or false?

You say or write a statement (such as '1 m 5 cm = 1.5 m', '9 is a factor of 3') and pupils decide if it is true or false.

Spider diagram

You write a whole number, fraction, decimal, percentage, word, algebraic expression … in a circle on the board. This is the spider's body. The 'legs' and 'feet' can be completed in a variety of ways.

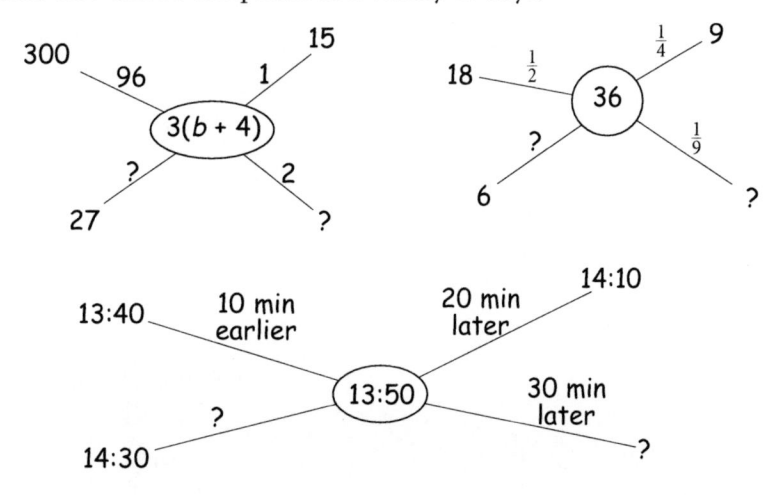

Mental methods and starters • 13

Today's number is …

Write a number on the board and put a ring round it. It could be a whole number (including negatives), decimal or fraction. Pupils make up calculations with that number as the result. Calculations can be restricted to a particular type. An example with the operations restricted to addition and subtraction is shown on page 21 of the pupils' book.

A variation of this format is *Today's expression is …*

What am I?

For example,
- I am a shape. I have three sides. Two of my sides are the same length. What am I?
- I am a two digit number. I am a multiple of 12. My nearest 10 is 50. What am I?

Ordering

Pupils put a set (usually numbers) in order. Examples are
- 5:15 p.m., 06:00, 15:30, 11:30 a.m., a quarter to four in the afternoon
- 0.3, 0.26, 0.7, 0.07
- 1.7 m, 118 cm, 1.03 m, 784 mm

Matching

Pupils match up sets into pairs or larger groups. Examples are

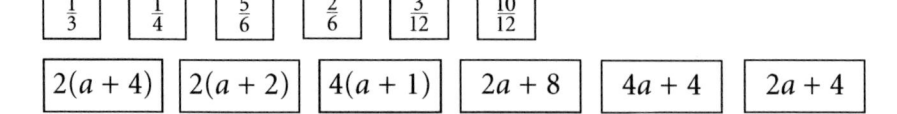

| $\frac{1}{3}$ | $\frac{1}{4}$ | $\frac{5}{6}$ | $\frac{2}{6}$ | $\frac{3}{12}$ | $\frac{10}{12}$ |

| $2(a + 4)$ | $2(a + 2)$ | $4(a + 1)$ | $2a + 8$ | $4a + 4$ | $2a + 4$ |

This format can be used to pair up the class for a further activity. Each pupil is given a card (showing for example, a whole number, fraction, expression, time, decimal, percentage …) and has to find the pupil whose card matches theirs in some way.

Target number

Pupils are given a set of numbers or choose them themselves.

A 'target' number is chosen. Pupils try to make the target number using the given numbers and any of the four operations. (Not all numbers need be used but each may be used only once.)

For example, make 474 from these.

        8   2   1   2   9   50

A possible solution is $(50 + 9) \times 8 + 2 = 474$

Bingo

Pupils choose numbers from a given set and fill in their bingo 'card' with them. The card can hold as many numbers as you choose.

For this card, numbers have been chosen from the set of integers ⁻10 to 10. You could 'call':

- ⁻2 + 8
- the number that is half way between ⁻3 and 1
- the temperature that is 2 degrees higher than ⁻6 °C

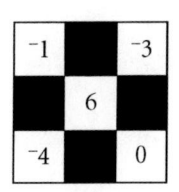

Counting stick

A counting stick is marked in equal intervals, usually in two contrasting colours. You can show pupils a starting point, tell them the value of one interval and then ask them to identify numbers you point to. For example, the value of one interval on this stick could be 2, 0.5, 0.1, 10 and the missing numbers identified in each case.

? 5 ? ? ? ?

Alternatively, give the values of two key points (Blu-tack labels on) and indicate points at random, each time asking for the value.

? 4 ? 5 ? ?

Odd one out

Pupils identify the odd one out from a set, such as

- $\frac{1}{4}$, 0.25, 0.14, 25%
- $\frac{1}{3}$, $\frac{2}{6}$, $\frac{3}{8}$, $\frac{5}{15}$

Array

This is a set of numbers, fractions, expressions … arranged in a grid or just in a list, on which you base questions, for example:

- Find two numbers that add to give ⁻1.
- Find three numbers that add to give 6.
- What is the total of the numbers in the second row?

⁻6	6	9
⁻5	8	⁻9
3	7	0

- Find two expressions that add to give $2n + 1$.
- Find two expressions that multiply to give $5n + 10$.
- Find three expressions that add to give $3n - 5$.

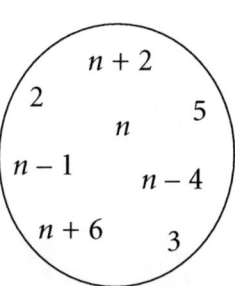

2, $n + 2$, 5, n, $n - 1$, $n - 4$, $n + 6$, 3

Mental methods and starters • 15

Two-way property table

The headings of a two-way table like this one are shown on the board and pupils say what could go in the cells. It is often best to say that all the entries should be different.

	A factor of 15	A factor of 12
A factor of 6	3	
A factor of 20		

Loop cards (I have … Who has …?)

These are commercially available but you can make up sets of your own. Each card has a question on it together with the answer to a question that is on another of the cards. The complete set forms a loop.

Give out the cards and decide who is going to start.

78	34 + 66		100	12 × 7

For example the pupil with the left-hand card above would start with 'Who has the answer to 34 + 66?' The pupil with the right-hand card would respond with 'I have 100. Who has the answer to 12 × 7?' If there are more pupils than cards, then some pupils can have two cards.

Around the world

One pupil is chosen to start who then stands behind the first pupil. Ask these two pupils a question. Whoever answers correctly first stands behind the next pupil. Carry on until each pupil has taken part. The aim is to end up furthest from your original seat.

Topics for starters

Below, arranged by some broad topic areas, are suggestions for other ways the starter formats can be used.

Integers

Array

Use a grid like this and questions such as these:
- Find two numbers that add to give 52.
- Find two numbers with a difference of 82.
- Find two numbers that multiply to give 72.
- What is the total of the numbers in the third column?

5	21	18
4	34	10
100	2	7

Using numbers from the loop, pupils find different ways to complete:

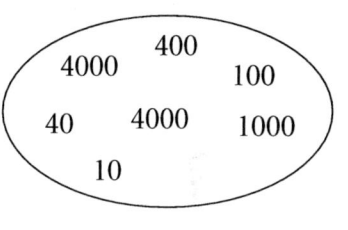

☐ ÷ ☐ = 40

☐ × ☐ = 40000

(Loop contains: 4000, 400, 100, 40, 4000, 1000, 10)

Pupils find as many different pairs of numbers as they can that
- add to give 3
- add to give 1
- subtract to give $^-2$

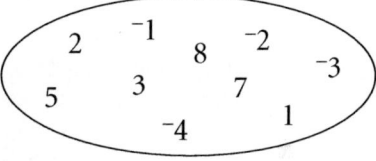

(Loop contains: 2, $^-1$, 8, $^-2$, 5, 3, 7, $^-3$, $^-4$, 1)

Bingo Use 'calls' like:
- 56 ÷ 8
- 42 ÷ 7
- 52 ÷ 4

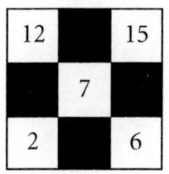

(Bingo card with: 12, 15, 7, 2, 6)

Ordering For example, 1009, 981, 909, 1200, 910

True or false For example, 30 × 50 = 150, $^-9\,°C$ is warmer than $^-5\,°C$

Counting stick Label it to include negative numbers.

Number relationships

Array Pupils use digits from the loop (at most once each) to make
- a two-digit multiple of 4
- a two-digit square number
- a cube number
- a two-digit prime number
- a three-digit number that is divisible by 3

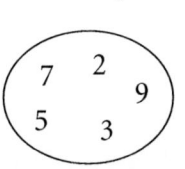

(Loop contains: 7, 2, 9, 5, 3)

What am I?
- I am a two-digit number. My digits are consecutive. I am prime. What could I be?
- I am a common factor of 36 and 24. I have two digits. What am I?

Odd one out Present sets such as 4, 12, 28, 23, 20 and 55, 80, 15, 30, 52

Two-way property table For example,

	Prime number	Square number
A factor of 8		
A factor of 7		

	Triangle number	A factor of 15
Prime number		
A multiple of 3		

Algebraic manipulation

True or false? For example, $n + n - 9 + 6 = 2n - 15$

Odd one out Pupils pick out the expression that isn't equivalent to the others from a set such as,
$n + 5 + n + 3$, $2(n + 4)$, $n + 10 + n - 2$, $2n - 2 + 6$

Matching A set of expressions has to be sorted into groups or pairs of equivalent expressions.

Bingo For this grid, numbers have been chosen from the integers 1 to 20.
Equations to be solved are called, such as
- $n + 3 = 18$
- $7n = 49$
- $3n - 2 = 4$

5		15
	7	
2		12

Fractions

True or false? For example, half this triangle is shaded (with an appropriate diagram), $\frac{1}{3} + \frac{1}{3} = \frac{2}{6}$

Spider diagram Put a fraction on the body and have that fraction of various amounts on the legs.

Ordering For example, a set of calculations such as $\frac{1}{2}$ of 30, $\frac{1}{3}$ of 36, $\frac{3}{4}$ of 24, $\frac{2}{5}$ of 40, $\frac{9}{11}$ of 33

Matching Match a set of 'fraction cards' with 'number cards' so that each pair gives the same result. For example,

| $\frac{1}{2}$ of | $\frac{3}{4}$ of | $\frac{2}{3}$ of | $\frac{6}{7}$ of | $\frac{4}{5}$ of | 14 | 18 | 16 | 15 | 24 |

Percentages

True or false? For example, 0.6 = 6%

Matching Pupils match a set of decimals with a set of percentages.

Ordering For example, 0.5, 43%, 0.1, 2%, 0.07

Bingo Questions called are 'What is 50% of 10?', 'What is 75% of 20?' and so on.

Decimals and measures

Array Use a grid like this and questions such as

- Find two numbers that add to give 5.
- Find two numbers with a difference of 4.9.
- Find two numbers that multiply to give 3.
- Find three numbers that add to give 3.
- What is the total of the numbers in the third row?

1	0.1	1.5
2.7	2	0.9
3.6	2.3	5

Spider diagram Put a decimal on the body and add and subtract amounts such as 0.1, 0.01 0.7 on the legs.

True or false? For example,

- 2.7 = 2.70
- 2.7 is bigger than 2.69
- 5.67 + 0.2 = 5.69

Matching Give a set of numbers written in different ways, such as 6.8, 6.08, 6.18, 6 + 8 tenths, 6 + 8 hundredths, 6 + 18 hundredths, 6 + 1 tenth + 8 hundredths

Bingo Calculations are called such as

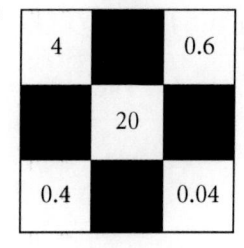

- 40 ÷ 100
- 0.04 × 100
- 4 ÷ 10

4		0.6
	20	
0.4		0.04

Ordering Give, for example,

- a set of rectangles to be sorted by area or perimeter
- a set of weights such as 6.4 kg, 200 g, 8000 g, 6500 g

Spatial visualisation

True or false For example,

- A triangle has at least one acute angle.
- A triangle can have side lengths of 8 cm, 1 cm and 3 cm.

What am I? I have 4 equal sides and rotational symmetry of order 2. What am I?

① First bites

7T/1, 7C/1

This is a collection of activities suitable for use in the first one or two weeks of year 7. Their purposes are

- to give pupils of all abilities an enjoyable and confident start to mathematics in the secondary school
- to give you a chance to get to know the pupils and how they work
- to help establish classroom routines and ways of working (whole class, group, individual)
- to give opportunities for homework

They do not not need a high level of number skill, so should be widely accessible.

There may not be time to do all the First bites activities at the beginning of year 7. Some may be left for later, perhaps as starting points for related units of work.

T	p 4	**A** Spot the mistake
T	p 4	**B** Four digits
T	p 4	**C** 4U + 1T
T	p 5	**D** Finding your way
T	p 6	**E** Gridlock
T	p 8	**F** Patterns from a hexagon
T	p 11	**G** Shapes on a dotty square

Essential	**Optional**
Sheets 45 and 46	Sheets 47, 48, 49 (blank grids) and 58
Up to 15 dice for a class of 30	Square and triangular dotty paper (sheets 1 and 2),
Sharp pencils, pairs of compasses, rulers,	3 by 3 pinboards, rubber bands, tracing paper,
coloured pencils, board compasses	OHP transparency of square dotty paper
Angle measurers	

A Spot the mistake (p 4)

> Sheets 45 and 46

These two resource sheets give pupils of all abilities an opportunity, in a light-hearted context, to spot some mathematical errors – and some non-mathematical ones! It is more fun for pupils to work in pairs, rather than on their own.

B Four digits (p 4)

This activity will tell you something about pupils' knowledge of number operations and symbols (for example, brackets) and their arithmetic skills. It also gives an opportunity for co-operative group work.

◊ Decide together on the four digits to be used. They do not have to be all different. 0 is not very helpful.

◊ You can allow free rein at first as far as the rules are concerned. Pupils will probably come up with some ground rules themselves, and then you can establish rules for everyone, for example:

- All four digits must be used.
- No digit can be repeated unless it occurs twice in the set.
- Digits can be used in any order.
- Any operations can be used. (Brackets may be needed and $\sqrt{}$ may be suggested by pupils.)
- Digits can be combined to make two- and three-digit numbers.
- Results must be whole numbers (for example, $43 \div (2 + 1) \neq 14$).

◊ You could start by asking for ways to make, for example, 10.

◊ Pupils could work in groups, each group making a collection. An element of competition could be introduced. Alternatively, groups could be given ranges of numbers (1–20, 21–40, …).

◊ It is necessary to record the completed numbers and the methods used to arrive at them, for example, a list:

1	11	21
2	12	22
3	13	etc.
4	14	
5	15	
6 1 + 3 + 4 – 2	16	
7	17	
8 2 + 3 + 4 – 1	18	
9	19 12 + 3 + 4	
10 1 + 2 + 3 + 4	20	

◊ In one school the results were recorded on a large chart and put on the wall. This was added to over the year. (It works particularly well if there is some reward for completing gaps – merits, credits, etc.)

◊ Calculators may be used if necessary, although many pupils should do well without them.

◊ Pupils may realise that results often come in pairs, for example:
  23 – 14 = 9, 23 + 14 = 37
then with some swapping around:
  32 – 14 = 18, 32 + 14 = 46, etc.

Follow-up

Each pupil can choose their own set of four numbers. Some sets of numbers (for example, 6, 7, 8, 9) are more difficult than others and may lead to demotivation.

ℂ **4U + 1T** (p 4)

◊ Start by asking for a two-digit number. (If 13, 26 or 39 are suggested, find a sneaky way of avoiding them!) Write the number on the board. Say that you are going to multiply the units digit by 4 and then add on the tens digit (so 37 becomes 4 × 7 + 3, giving 31). Write the result on the board.

Ask pupils to do the same to this new number. Write the result. Repeat a few times until all pupils have understood the rule for generating the next number. Do not go on too long or you will form a loop. It is best for pupils to discover the loop for themselves.

You may come across a single-digit number in the process. If not, you should introduce one and discuss how to deal with it.

Ask pupils each to choose a two-digit number of their own, use the rule to make a chain of numbers and watch what happens as their chain grows.

After a while, someone will notice they have come back to a number they had before. When this happens, discuss it with the class. Pupils should realise that once a loop is formed, no new numbers are generated, but they do not always find this obvious!

◊ Questions for investigation are:
- Will all numbers form loops?
- How long are the loops?
- How many different loops are there?
- Do any numbers go straight to themselves? (Yes: 13, 26 and 39!)

There is nothing special about 4 as the multiplier: it just gives short chains. Pupils can investigate other multipliers. How many chains are there in each case? (The rule 2U + 1T produces a nice overall diagram with every number connected to one big loop, except for the multiples of 19.) More able pupils can be challenged to work backwards. For example, if the rule is 2U + 1T, what numbers generate 15 as their next number?

D **Finding your way** (p 5)

This gives practice in using left and right and in reading a simple map.

◊ Before discussing the picture and the questions, you could use a plan of your school. Give pupils a list of instructions from the classroom to somewhere else and ask them to guess the destination. They can then make up instructions for one another.

You could ask pupils to shut their eyes and imagine where they are going as you give them instructions.

Similar work can be based on local maps

E **Gridlock** (p 6)

This game gives you an opportunity to find out pupils' addition (and subtraction) skills. Pupils can also develop and explain strategies to win.

> Up to 15 dice for a class of 30
> Optional: Sheets 47, 48 and 49 (blank grids)

◊ To help them understand the scoring system, pupils could complete the grids on sheet 47 and work out the scores. Some schools have used this sheet for homework.

'I almost didn't use sheet 47 but it turned out to be most useful.'

◊ Initially the class could play together, with you rolling the dice and calling the numbers. Then the game can be played in groups of two or more.

1 First bites • 23

◊ More able pupils may think that the game is a trivial exercise requiring only simple addition skills. Emphasise early on that they should be thinking about good strategies to maximise their chance of winning.

> **To play 'Gridlock'**
>
> Each pupil draws a square grid (start with 3 by 3 grids) and marks off the top left-hand section as shown.
>
> The caller rolls a dice and calls out the number. Each pupil writes the number in any empty square in the section shown shaded on the right.
>
> Repeat until each square in that section is filled.
>
> Each number must be written in the grid before the next is called and a number can't be changed once it is written.
>
> Each pupil adds up their numbers in the rows, columns and diagonal and writes the totals in the empty squares as shown on the right.
>
> Each pupil adds up their points.
>
> • Score 2 points for a total that appears twice.
>
> • Score 3 points for a total that appears three times.
>
> • Score 4 points for a total that appears four times ... and so on.
>
> The grid above scores 4 points (6 and 7 both appear twice as a total).
>
> After a number of rounds (decided by you), pupils add up their points and the one with most points is the winner.

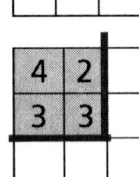

◊ After playing on 3 by 3 grids, play the game on larger ones.

◊ Once pupils have played the game a few times, ask them to describe any strategies they use in placing the numbers on their grids. For example, if a number is rolled twice it is better to place the numbers diagonally, for example [5 _ / _ 5] or [_ 5 / 5 _] rather than [5 5 / _ _]

'I tried moving on to 5 by 5 grids and it was useful. I rolled 16 dice, wrote the numbers on the board and challenged pupils to achieve 9 points and 0 points. This exposed the value of the diagonal symmetry ... Pupils greatly enjoyed this and preferred this variation to the original game.'

◊ Now you can alter the rules as follows. First, the numbers called out are written at the side of the grid. When all numbers have been called, they are then placed in the grid. Pupils can think about how to get the maximum possible score with a particular set of numbers.

◊ One variation is for the winner to be the person with the fewest points. Pupils can discuss how their winning strategies change in this case.

Another variation is to use two dice to generate larger numbers.

Follow-up

In E1 to E11, the later questions are more difficult.

Remind pupils that they can only use numbers on an ordinary dice (1 to 6) to solve these problems.

E1 In (b), emphasise that their problems should be able to be solved without any guesswork or mind reading! Encourage more able pupils to make up problems that give the minimum necessary information.

This could be set as a homework task.

◊ You could ask pupils how changing the system of scoring points would affect the game, for example:
- score 2 points for a total that appears twice
- score 4 points for a total that appears three times
- score 6 points for a total that appears four times ... and so on

◊ A different version of the game is for pupils to cross out any totals that repeat and to add the remaining totals to give their score for that round. The winner could be the person with the most or fewest points.

F Patterns from a hexagon (p 8)

This work is to help pupils develop skills with compasses and rulers that are needed later to construct triangles and angles. Pupils also analyse patterns and make decisions about how to construct them.

At the start of the year many pupils have coloured pencils and geometry sets, so capitalise on this.

> Sharp pencils, pairs of compasses, rulers, coloured pencils, board compasses

'I discovered that only about half the class had used compasses before.'

◊ Many pupils find it difficult to draw a circle with a pair of compasses. They may need to draw circles and simple patterns before they feel confident enough to try the more difficult patterns.

Many pupils will find it helpful to see a demonstration of how to draw a regular hexagon. They must be able to draw a regular hexagon in order to draw the patterns on page 9.

Common problems include:
- not realising that the point of the compasses is moved to the point where the last arc crosses the circumference (and not at the end of the arc) for subsequent arcs to be drawn
- not realising that, to draw the hexagon, you join points where the arcs cross the circumference (and not the ends of the arcs)

1 First bites • 25

Although pupils may have drawn them before, it may be helpful to demonstrate on the board or OHP how to draw the seven-circle or petal designs shown below.

◊ This work provides good material for wall displays. In one school, the hexagon designs were used to make mobiles.

◊ You may want pupils to leave construction lines so you can check their methods.

◊ The construction of the patterns on page 9 offers more of a challenge, and the construction gets more involved further down the page.

The last two designs can be drawn as follows:

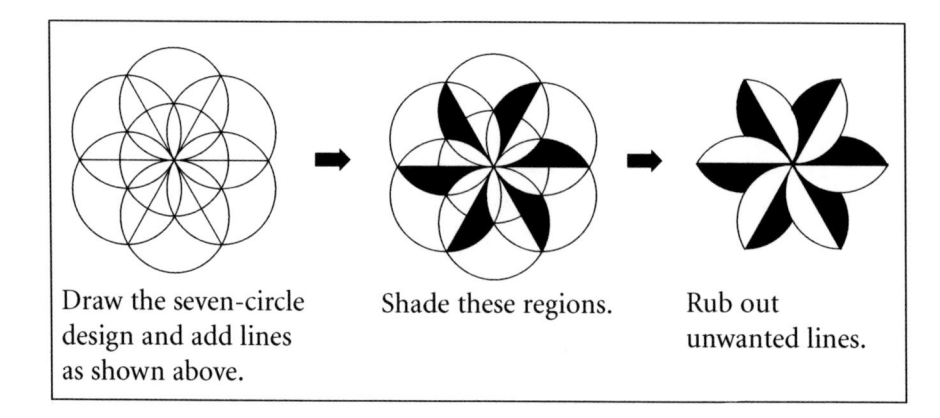

Draw the seven-circle design and add lines as shown above. → Shade these regions. → Rub out unwanted lines.

Mark six equidistant points round a circle.

Draw a circle, centre O and radius OP.

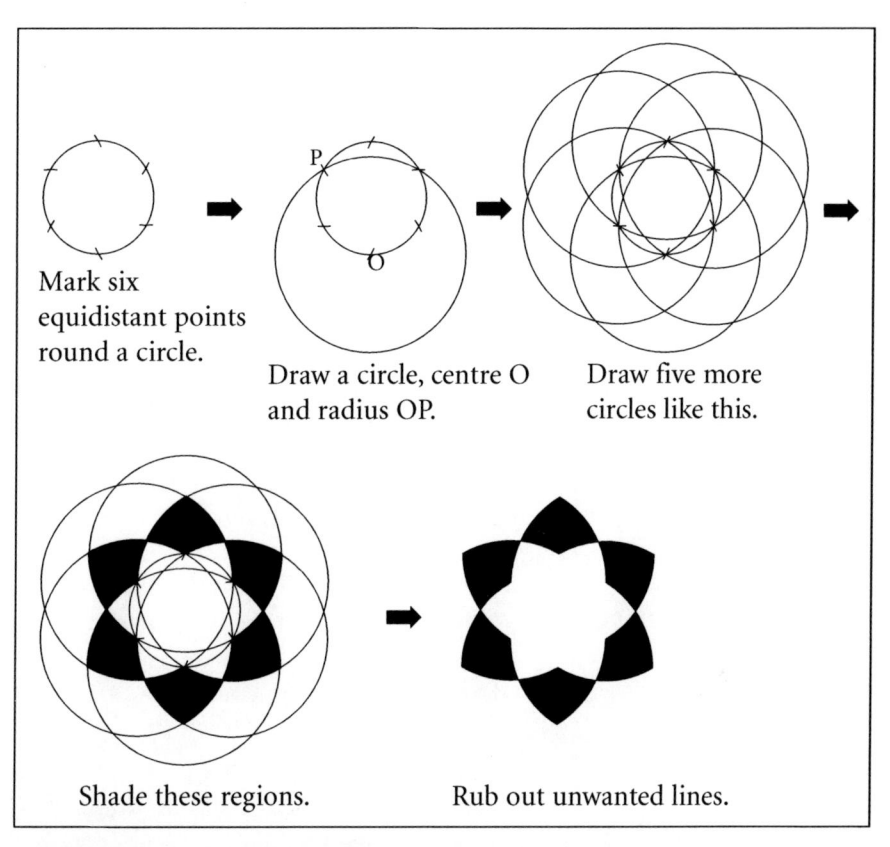

Draw five more circles like this.

Shade these regions.

Rub out unwanted lines.

Follow-up

If they know about symmetry, pupils could try to draw a pattern with 0 lines of symmetry, 1 line of symmetry etc.

G **Shapes on a dotty square** (p 11)

Pupils create shapes and use mathematical language to describe them. They can also decide on their own lines of investigation.

> Optional: Square and triangular dotty paper (sheets 1 and 2), sheet 58, 3 by 3 pinboards, rubber bands, tracing paper, OHP transparency of square dotty paper

◊ Square dotty paper can be used or pupils can draw grids of dots on square paper. Sheet 58 has the grids already ruled off.

◊ Establish rules for drawing shapes on the pinboard/grid.
- Only the 9 pins/dots can be used.
- All corners must be at a pin/dot.
- The types of shape shown in the pupil's book are disallowed (ones with 'crossovers' or 'sticking out' lines).

'Computers helped resolve arguments as to "sameness" of shapes by rotating, reflecting and superimposing.'

◊ Ask pupils to draw a few different shapes following the rules above.

There is likely to be some discussion on the possible meanings of 'different' and 'same' here. 'Different' is usually taken to mean non-congruent. Tracing paper helps pupils identify shapes that are the same.

Look at some of their shapes together.

- What properties have they got?
  (Number of sides, angles, symmetry, parallel sides, area, ...)
- Do pupils know names for any of the shapes?
  (Triangle, quadrilateral, rectangle, parallelogram, hexagon, ...)

◊ There are various ways to structure this activity. Some trial schools generated a collection of questions from which pupils chose. For example:

What shapes have the most sides?
How many different squares?
How many different triangles?
How many shapes have reflection symmetry?
How many shapes have rotation symmetry?
What different areas can you make?
How many ways can you put the same triangle on the grid?
What shapes can be made with 1, 2, 3, ... right angles?

Pupils can choose a question (or pose one of their own) and write up their solution. These could then be displayed.

For example, the 23 different polygons with reflection symmetry are:

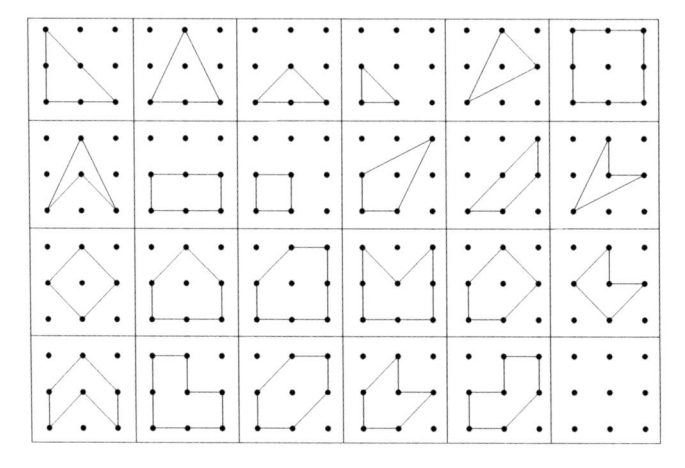

28 • 1 *First bites*

◊ An alternative structure is to begin by asking how many different triangles can be found. The 8 different triangles are:

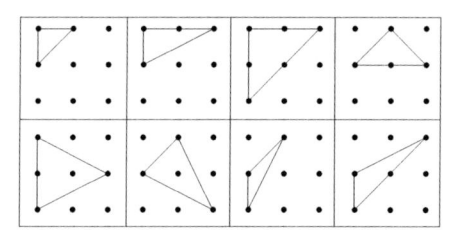

Now discuss the properties of the triangles. For example:
  Which has the greatest area?
  Which has a right angle?
  Which has reflection symmetry?
  Which are isosceles?

Pupils can now consider the different quadrilaterals that can be found and their properties. The 16 different quadrilaterals are:

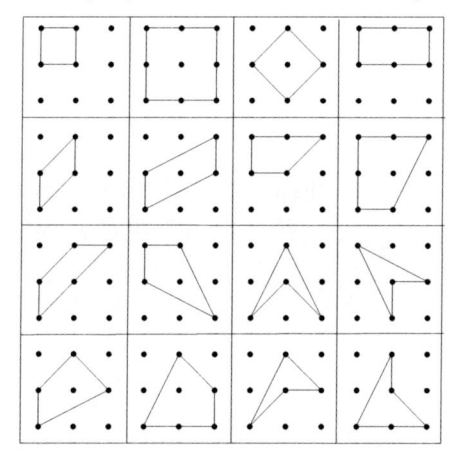

'I put the class into teams, following a lesson and a homework, to find all possible polygons, to collate results and present.'

Pentagons, hexagons and heptagons can be considered but the number of different shapes may be rather daunting!

The numbers of different polygons of each type are:

Number of sides	Number of different polygons
3	8
4	16
5	23
6	22
7	5

This gives 74 different polygons.

◊ One extension is to consider polygons on a different grid. Suggestions appear in the pupil's book.

The hexagonal 7-pin grid yields 19 different polygons which is a manageable number for most pupils to find. The polygons are:

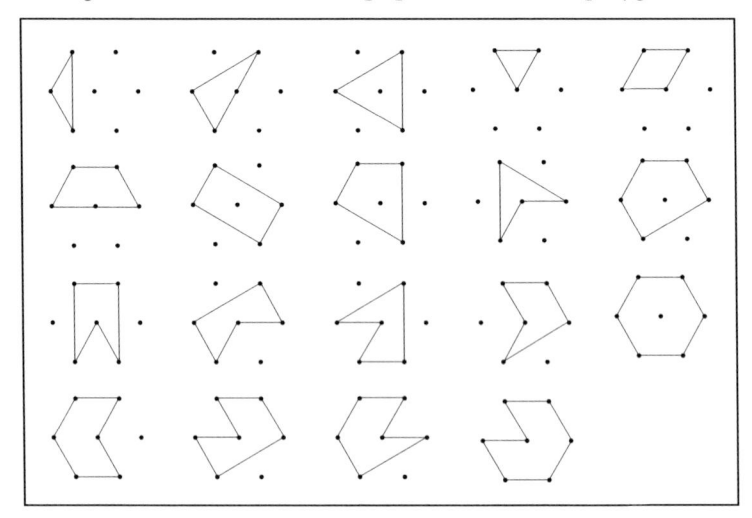

A Spot the mistake (p 4)

1 Off to Benidorm in June

Fish tank framework is an impossible object.
Table left-hand rear leg is longer than the others.
Vase on the table is an impossible object.
Sofa is an impossible object.
Socket on left-hand wall has holes in wrong orientation.
Mirror reflection is incorrect; 'TAXI' and the clock face are the wrong way round.
Front door has handle and hinges on the same side.
The '33' above the front door is the wrong way round.
Vacuum cleaner plug has only one pin.
Vacuum cleaner hose has an extra hose tangled in it.
Triangles on shelf: right-hand one is an impossible object.
View of window blind is impossible.

2 Taxi to the airport

Clock has a back-to-front 3, and 8 where it should be 9.
P (parking) sign is on a road with double yellow lines.
Stop sign on road – S is wrong way round.
The word STOP (and the road marking) is on the wrong side of the road.
Bicycle has no front wheel.
Airport sign says 5 cm (centimetres).
Traffic lights have a right turn arrow to a no-entry street.
Rollerblader has one ice skate on.
Right-hand no-entry sign is pointing upwards.
Pedestrian crossing markings on the road should be rectangles.
Pedestrian crossing beacon is the wrong shape.
Street lamp on top of the no-entry sign is facing up.
A vegetarian butcher would not do much business!

Low bridge sign says min(imum) and should say max(imum).

Taxi is going to the airport, so should have turned left.

('Tax to rise 150%' is not necessarily wrong – it could!)

3 At the airport

A plane is flying upside down.
The plane taking off has no tail wings.
Passengers are walking along the wing of the waiting plane.
Waiting plane has RAF insignia on tail.
Waiting plane has a ski instead of a wheel.
Wind socks are blowing in opposite directions.
Tannoy message says 'train' instead of 'plane'.
Tannoy message contradicts time on clock.
Christmas tree contradicts Easter eggs sign.
Tax free sign: you cannot save 200%.
Suitcases are ticketed to Rome, and flight is to Benidorm.
Suitcase on weight machine has a square wheel.
Luggage weight sign says WAIT, not WEIGHT.
Luggage weight is in g(rams) and should be kg (kilograms).
The 'All departures' sign points to a no entry corridor.
You cannot 'Ski the Pyramids'.
You cannot 'Ice skate the Amazon'.

4 The hotel reception

Clock reads 14:60, and the minutes must be less than 60.
Sign above toilet doors says 'Welcome to BeMidorN.'
Calendar on reception desk says 31 June (only 30 days in June).
A Christmas tree in June is rare!
Toilet door pictures are the wrong way round.
Right-hand toilet door has handle and hinges on the same side.
Change sign – 100 pts should be 1000 pts, and the peseta has of course been replaced by the euro.
Double rooms are cheaper than single rooms.
Atlantic views should read Mediterranean views in Benidorm.
Left-hand rear leg on table is longer than other three.
Tickets on luggage have changed since the airport to Roma and Home.
Plant doesn't sit in its flower pot.
Lift is on ground floor (the lowest in the list above the door) but shows on the list (and the call buttons) as going further down.
List of floors above lift door is missing floor 4.
Sign in lift mirror – S is wrong in reflection.
Sign in lift has a weight limit that is silly.
(The vase on the table is not an impossible object!)

5 By the pool

You don't get whales in the Mediterranean.
The flag at the top of the boat's mast should be blowing forwards.
Speed boat and water skier are not connected.
Plane is flying backwards if it is pulling the banner.
Banner should read 24 hours not 26.
Weather vane NSEW are wrong.
If the time is 23:30 the sun would not be out.
Temperature of $^-26°C$ is a bit chilly for sunbathing.
Water level in man's jug should be horizontal.
Sun lounger nearest to front is missing a leg.

Gazebo on the left-hand corner of the balcony is an impossible object.
Diving board heights are in millimetres, and should be in metres.
Lower diving board is at a greater height (8) than the upper (4).
There appears no means of access to the lower diving board.
'Do not feed fish' sign is unlikely in a swimming pool.
Man fishing not possible (we hope) in a swimming pool.
Swimming pool depth signs are wrong.
Shark in swimming pool.
Stairs and railings up to balcony create an impossible object.
Sangria jugs of 75 l(itres) would be a bit big!
Children shown standing at the left-hand side of the pool where depth is 5 m(etres).
Depth shown as 10 cm where the diving boards are.

D Finding your way (p 5)

D1 Robin Hall

D2 The pupil's journey

E Gridlock (p 6)

E1 (a)

2	3	5
3	4	7
5	7	6

Points scored: **4**

4	1	5
3	6	9
7	7	10

Points scored: **2**

1	6	5	12
5	3	1	9
6	4	6	16
12	13	12	10

Points scored: **3**

(b) The pupil's problems

E2 Examples of grids that score 2 points are:

5	3	8
6	2	8
11	5	7

3	6	9
5	2	7
8	8	5

E3 Examples are:

(a)

6	6	12
4	3	7
10	9	9

(b)

4	6	10
6	3	9
10	9	7

E4 (a) 2 points (b) 0 points

E5 (a)

4	2	6
2	1	3
6	3	5

(b)

3	1	4
2	5	7
5	6	8

(c)

5	2	3	10
4	6	4	14
1	1	1	3
10	9	8	12

E6 Examples are: 1, 2, 3 and 5; 2, 3, 4 and 6.

E7 The pupil's explanation

E8 The pupil's explanation

E9 (a)

6	4	10
4	3	7
10	7	**9**

(b)

1	**5**	6
4	**6**	10
5	11	7

E10 Examples are:

(a)

1	2	6	9
2	3	5	10
6	5	4	15
9	10	15	8

(b)

1	2	5	8
2	3	6	11
5	6	4	15
8	11	15	8

(c)

5	1	2	8
5	2	3	10
4	6	6	16
14	9	11	13

E11

1	**5**	**5**	11
1	**3**	**4**	8
6	**5**	**6**	17
8	13	15	10

Sheet 47

1.

6	5	**11**
1	1	**2**
7	**6**	**7**

Points scored: **2**

2.

3	4	**7**
6	5	**11**
9	**9**	**8**

Points scored: **2**

3.

3	5	**8**
2	4	**6**
5	**9**	**7**

Points scored: **0**

4.

3	4	**7**
4	3	**7**
7	**7**	**6**

Points scored: **4**

5.

6	1	2	**9**
3	4	6	**13**
1	5	1	**7**
10	**10**	**9**	**11**

Points scored: **4**

6.

5	3	5	**13**
3	1	4	**8**
2	6	1	**9**
10	**10**	**10**	**7**

Points scored: **3**

② Symmetry 1

7T/5, 7C/2

Pupils use folding, cutting and mirrors in a variety of practical activities. They have especially enjoyed the symmetry tiles game (section D) and it has proved effective in consolidating the key ideas.

Rotation symmetry is not dealt with in this unit. However some pupils are likely to be aware of the idea and it acts as a 'distractor' in some questions. You may wish to discuss it where appropriate.

T	p 12 **A** Finding mirror images	Predicting a mirror image (including sloping mirror lines) – checking by folding and cutting or using a mirror
	p 13 **B** Identifying reflection symmetry	
T	p 14 **C** More than one line of symmetry	Folding and cutting Line symmetry in letters of the alphabet
	p 15 **D** Symmetry tiles	
	p 18 **E** Making patterns	
	p 19 **F** Shading squares	

Essential	**Optional**
Mirrors Scissors Paper for folding and cutting Square dotty paper or centimetre squared paper Sheets 59 to 66	Tracing paper

Practice booklet pages 4 to 6

Ⓐ Finding mirror images (p 12)

> Scissors, mirrors, sheets 59 to 63
> Optional: Tracing paper

T ◊ For the folding and cutting exercise, pupils could use alternative methods to check (a mirror and/or tracing paper) but folding and cutting gives an immediate check that is very convincing.

34 • 2 Symmetry 1

> 'As an introduction, pupils drew the 2nd half of a mask. They found it fun and the masks made a good display. They knew what to do in a non-mathematical context. This built confidence.'

◊ You could miss out some of sheets 59 to 62 depending on pupils' prior attainment. Sheet 61 should reveal the common errors when reflecting in sloping lines. Counting dots outwards from the mirror line can be a useful approach.

◊ It may be worth leading the class through the stages in questions A3 and A4, stressing the meaning of the word 'image'.

B Identifying reflection symmetry (p 13)

Pupils identify lines of symmetry and use a mirror to check.

> Mirrors

◊ It may be worth leading the class through the stages in the photographs.

B2 Some pupils may say that design (d) has reflection symmetry because they recognise the rotation symmetry. It is a good opportunity to discuss this type of symmetry.

C More than one line of symmetry (p 14)

Pupils predict and make shapes with two lines of symmetry by folding and cutting. They also find lines of symmetry for the letters of the alphabet.

> Scissors, loose sheets of paper (newspaper might be useful), sheet 66
> You could prepare some shapes on 'twice-folded paper' such as the ones at the top of page 14 to use in your introduction.

◊ One way to structure the introduction is as follows.

Fold a sheet of A4 in half and half again like this.
Make sure pupils can see where the folds are.
(Draw along each fold with felt-tip pen if there is any doubt.)

Draw on the paper, for example like this.
Ask what the shape will be like when you cut it out and unfold it. Unfold the sheet for pupils to see.

Now show the three examples at the top of page 14.

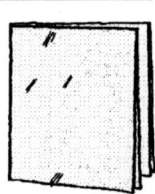

> 'In C3, some cheated by trying to cut shapes and then fold! Lots of teacher input ... but a good activity for identifying weaknesses and developing ideas.'

Ask pupils which of them will make a letter of the alphabet when cut and opened out. Test their prediction by cutting and opening out.

Ask pupils to describe what the other shapes will look like when they are cut and opened out. Test their predictions by cutting and opening out.

C4 The 'O' on sheet 66 is a circle and so has an infinite number of lines of symmetry.

D **Symmetry tiles** (p 15)

Pupils consolidate work on reflection symmetry by using tiles to make symmetrical patterns. The tiles are also used to play a game.

> Scissors, mirrors, sheet 64 (copied on card if possible),
> sheet 65 (one for each group of players)

◊ In problems D1 to D5 there is no need for pupils to draw diagrams to show their results but some may wish to do so.

Symmetry tiles game

◊ The game should be self-correcting, with players protesting at an invalid move. Some care is needed in organising the groups: each group should have one pupil who is confident enough about symmetry to recognise invalid moves. Although the game can be played with four players, having only two or three makes it faster and more enjoyable.

◊ You may have to clarify one or two things: pupils can put their cards down on either side of the dotted line; they don't have to complete the symmetry at every move (though they could play that way).

◊ Watch for pupils who are still getting the symmetry wrong. Provide them with a mirror so they can see their mistakes.

E **Making patterns** (p 18)

Pupils make patterns with four lines of symmetry.

> Squared or dotty paper
> Optional: Mirrors or tracing paper

◊ Remind pupils that any colouring should be symmetrical too.

F Shading squares (p 19)

Pupils make symmetrical designs by shading squares. They can try to devise strategies to ensure all the different ways of shading squares are included.

Optional: Mirrors (to check results)

◊ Encourage pupils to explain how they know they have found all the different ways for each problem.

◊ These problems can be extended in a variety of ways. In one school, pupils suggested their own extensions and looked at shading different numbers of squares on these diagrams.

A Finding mirror images (p 12)

A1 to A2 The pupil's drawings and checks

A3 to A4 The pupil's symmetrical drawings

B Identifying reflection symmetry (p 13)

B1 (b) Yes   (c) Yes   (d) No
      (e) No   (f) Yes

B2 (a) Yes (b) No (c) Yes (d) No

C More than one line of symmetry (p 14)

C1 Shape R

C2 Shape B makes I and shape C makes X.

C3 The pupil's attempts to produce the shapes

C4 (a) The first letter with no line of symmetry is F.
      (b) A few lines of symmetry are drawn on 'O' as examples: 'O' has an infinite number of lines of symmetry.

(c)

Number of lines of reflection symmetry	Letters
0	F G J L N P R S Z
1	A B C D E K M Q T U V W Y
2	H I
3	
4	X
more than 4	O

D Symmetry tiles (p 15)

D1 The pupil's symmetrical patterns

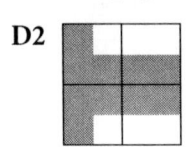

D2

D3 In each case, there is another solution with the tiles on the other side of the dotted line.
(a) (b)

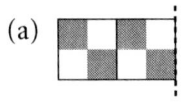

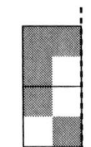

D4

D5 In each case, there is another solution with the tiles on the other side of the dotted line.
(a) (b)

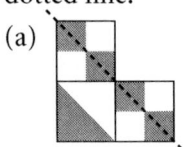

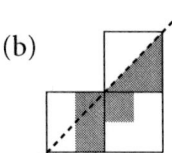

D6 In each case, there is another solution with the tiles on the other side of the dotted line.
(a) (b)
(c) (d)
(e) (f)
(g) (h)

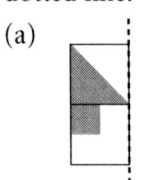

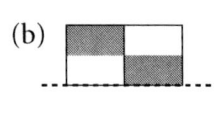

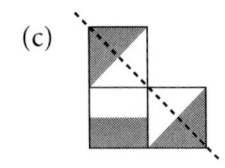

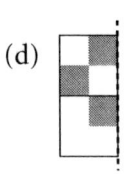

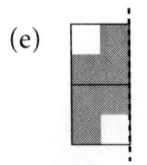

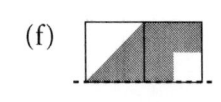

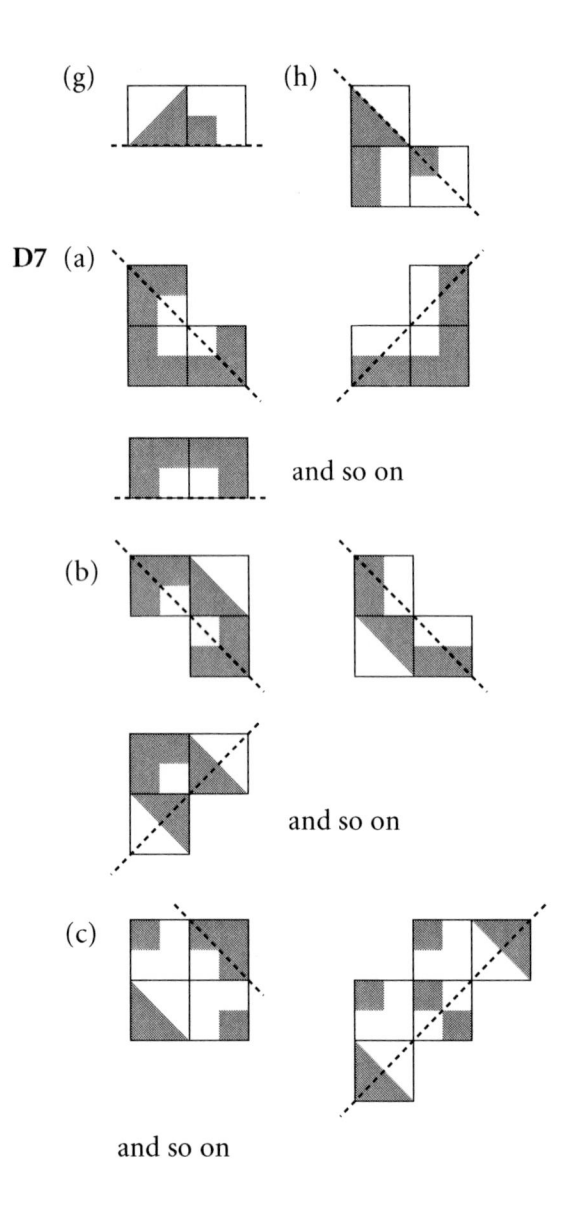

D7 (a)

and so on

(b)

and so on

(c)

and so on

E Making patterns (p 18)

E1 (a) Two lines of symmetry
(b) Four lines of symmetry

E2 The pupil's drawings

F Shading squares (p 19)

F1 Ten different ways

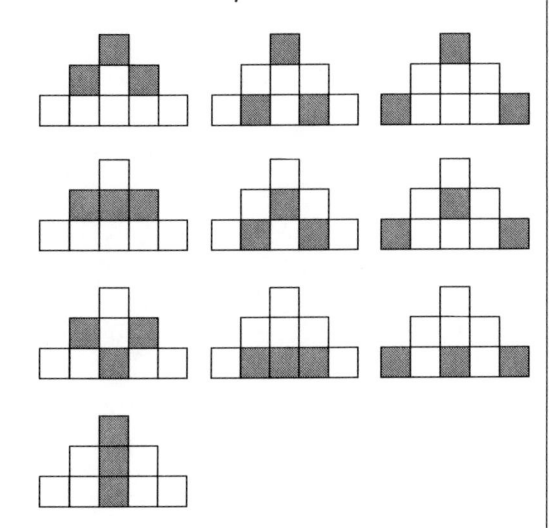

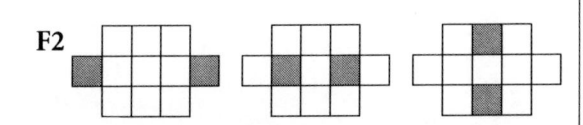

F2

What progress have you made? (p 19)

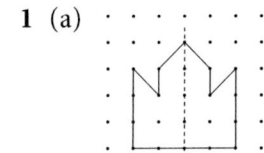

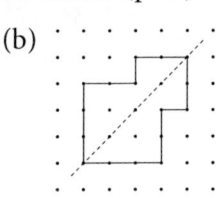

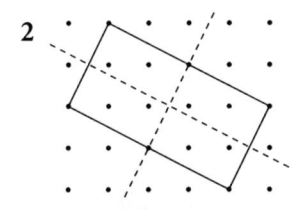

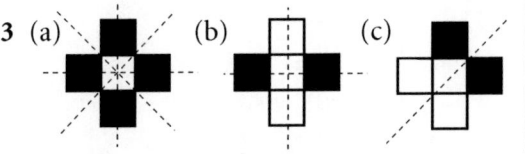

1 (a) (b)

2

3 (a) (b) (c)

Practice booklet

Section A (p 4)

1 The pupil's drawings

Section B (p 5)

1 (a) No   (b) No   (c) Yes

2 (a) Yes (b) Yes (c) Yes (d) No
  (e) Yes (f) No   (g) No   (h) Yes
  (i) Yes

Sections C, D and E (p 6)

1 (a) 3   (b) 2   (c) 4

2 4

3 The pupil's drawings

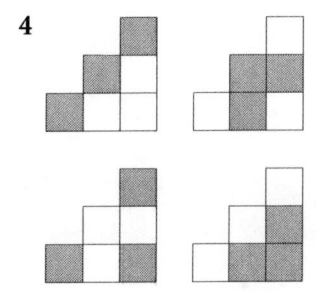

4

3 Whole number calculation 7T/3, 7T/6, 7T/21, 7C/3

T	p 20	**A** Adding and subtracting – mental methods
T	p 22	**B** Adding and subtracting – written methods
T	p 26	**C** Up to 10 × 10
T	p 29	**D** Division
T	p 31	**E** Multiplying by a single-digit number
T	p 33	**F** Dividing by a single-digit number
T	p 35	**G** What do we do about remainders?

Essential	**Optional**
Sheets 19, 35	Dienes base ten equipment/10p and 1p coins
	Sheets 25, 36
	Dice (a variety of types)
Practice booklet pages 7 to 10	

A Adding and subtracting – mental methods (p 20)

> 'My students were very possessive of their own methods!'

This work will give pupils the opportunity to see mental methods of calculation other than their own, and to see that their own idiosyncratic methods may not be unusual. The activity is not designed to impose methods on pupils. Class discussion is important, but pupils are likely to be more forthcoming in describing their methods in pairs or small groups first.

> 'All these methods [on page 20] arose in discussion and proved useful for those who didn't have an efficient method.'

◊ Present some calculations in context, for example: 'You need £55 to buy a Walkman. You have saved up £28. How much more do you need?' Emphasise it is their methods you are interested in, not just the answer. It is better to discuss several different methods for each of a few calculations than to have one or two methods for lots of questions.

◊ The number line is good for visualising addition and subtraction. The diagrams in the pupil's book illustrate some ways of doing calculations. Pupils need to realise that no method is more correct than any other.

◊ Further guidance on addition and subtraction is given on pages 7 to 10 in the section on mental methods.

Today's number is ...

◊ See page 14 of this guide for a description of 'Today's number is ...'.

◊ Pupils could be restricted to addition and subtraction here.
They could be allowed to choose the number for themselves.

B Adding and subtracting – written methods (p 22)

You may wish pupils to do the addition practice on page 23 before they play 'Total'.

Total

> A set of cards numbered 0 to 9 (sheet 19) or a dice (for each group)

◊ After a few games discuss any strategies pupils use in positioning the digits. If pupils are using dice instead of cards, discuss ways in which they would change their strategies if they used a different dice. For example, using a 1 to 6 dice, if 5 was the first digit rolled where would they put it? If they used a 1 to 8 dice would they put a 5 in the same position in their grid?

◊ Pupils could play the game again but change the rules in some way.
For example, the winner could be the player
  - with the lowest score
  - whose score is nearest an agreed target number

◊ Ask pupils to make 1000 (or 100 at first) by choosing any of the cards.
After some time discuss any strategies they used to find a solution.
In particular discuss why the digits in the units column must add up to 10, and the digits in others must sum to 9. (There are 192 different solutions for 1000, so do not expect them to find them all!)

Written subtraction practice

Those who find written subtraction difficult may need individual help.

> Optional: Dienes base ten equipment or 10p and 1p coins,
> a set of place value cards (such as those on sheet 25)

◊ The term 'borrowing' is misleading. It gives the the impression of an unfinished transaction. A more appropriate term is 'exchanging': we exchange a ten for 10 units or a hundred for 10 tens.

◊ Point out that it is a good idea to check the answer to a subtraction by adding. For example, the calculation on page 23 can be checked by showing that 128 + 217 = 345.

◊ Coins can be used in two-digit subtraction. For example, 35 – 17 can be tackled by first showing that 3 tens and 5 units can be exchanged for 2 tens and 15 units.

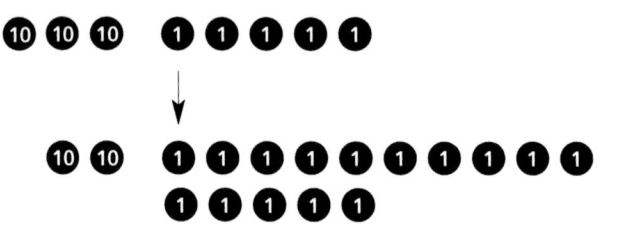

⑩ ⑩ ⑩   ① ① ① ① ①
        ↓
⑩ ⑩   ① ① ① ① ① ① ① ① ① ①
       ① ① ① ① ①

Now subtract 17 to leave 18.

⑩   ① ① ① ① ① ① ① ①

◊ Place value cards show how a three-digit number like 645 is made up of 600 + 40 + 5. Using them can help pupils make sense of the subtraction process. For example, 645 – 268 can be done like this.

$$\begin{array}{r}645\\-268\end{array} \qquad \begin{array}{rrr}600 & 40 & 5\\\underline{200} & \underline{60} & \underline{8}\end{array}$$

$$\begin{array}{rrr}600 & 30 & 15\\\underline{200} & \underline{60} & \underline{8}\\ & & \underline{7}\end{array} \quad \text{(exchanging 10)}$$

$$\begin{array}{rrr}500 & 130 & 15\\200 & 60 & 8\\\underline{300} & \underline{70} & \underline{7}\end{array} \quad \text{(exchanging 100)}$$

Result: 377

Number magic

◊ All pupils can try the investigation on two-digit numbers.

Extending to three or four digits may well prove too much for pupils whose written subtraction is weak unless they are able to use a calculator.

With two-digit numbers pupils may notice that

  • The first answer is always in the 9 times table.
  • The final answer is always 99 (treating 9 as 09).

With three-digit numbers the final answer is always 1089 (treating 99 as 099).

With four-digit numbers the final answer depends on the middle pair of digits. If the 2nd digit is greater than the 3rd, the result is 10890. If the 2nd and 3rd digits are equal, the result is 10989. If the 2nd digit is less than the third, the result is 9999.

Encourage more able pupils to persist and try to find a pattern in their results.

Largest and smallest

◊ The process always ends up with the number 6174.

Target 1000

'I set this as homework. Pupils enjoyed it and a lot got their parents to help.'

◊ A total of 1000 is impossible, but there are very many ways of getting a total of 999. The units column has to add to 19, the tens column to 18 and the hundreds column to 8. For example, 79 + 84 + 516 + 320.

C **Up to 10 × 10** (p 26)

> Sheet 35
> Optional: dice numbered from 4 to 9, sheet 36

◊ You could begin the discussion by asking pupils which tables facts they think they know 'by heart' and which they have to think about. Discuss how they work these out. A variety of methods of calculating 6 × 8 are shown and these could support your discussion. Most rely on knowledge of 'easier' tables facts.

Discuss quick methods for other multiplications.
Some possibilities are
 • to multiply by 5, multiply by 10 and then halve the result
 • to multiply by 4, double and then double again
 • to multiply by 6, multiply by 3 and then double the result
 • to multiply by 9, multiply by 10 and then subtract

Pupils may be interested in looking at patterns in the 9 times table and this could help their recall.

$9 \times 1 = 9$
$9 \times 2 = 18$
$9 \times 3 = 27$
$9 \times 4 = 36$
$9 \times 5 = 45$ ...

Pupils may be able to use the fact that the first digit of $9 \times n$ is always one less than n and the digits of $9n$ add up to 9 (for $n = 1$ to 10) to help them learn these facts.

◊ Each pupil could be issued with a standard 'tables square' (or write out their own). They could put stickers on it to cover the tables facts they know already and add to it as they learn more. The tables square can be used to show that the number of facts to memorise is less than might be thought at first: if you can put a sticker over, say, 3 × 7 then 7 × 3 can be covered as well.

◊ Pupils often enjoy 'Tables bingo'. The caller could use two dice numbered 4 to 9. Write a list of, say, 20 numbers on the board, from which each pupil picks seven to be their 'card'. (Include some non-starters, like 47, and ask later why they are avoided.) The caller rolls the two dice to generate two numbers to be multiplied and each pupil crosses out the product if it appears on their card. The first person to cross out all the numbers on their card and call 'bingo' is of course the winner.

◊ Instead of the standard 'tables test', you could use a pack of ordinary playing cards (without picture cards) to generate the numbers to be multiplied. The class will feel smug when they get 1 × 2, etc.!

◊ More consolidation activities are described below.

Grids

◊ In many trial schools, this activity was found to be popular and useful. One school commented that it could be suggested to parents as a way of practising tables with their children.

Mixed-up tables

> Optional: Sheet 36 (to save drawing grids)

◊ Pupils usually find the race very motivating!

Cover up

> Sheet 35

◊ Many trial schools found this a useful homework activity.

Pairs

◊ To get the largest result, you pair off the two largest numbers, then the other two numbers.

The mixed-up table puzzle

Schools have commented on how much pupils enjoyed cracking the code and explaining how they had done it.

D Division (p 29)

In this section divisions are restricted to those where the number to be divided does not exceed 100.

> Optional: Sheet 36 (for D3)

◊ You could begin the discussion by asking pupils to solve pairs of problems such as 'Share 12 cakes equally between 3 people' and 'If 12 cakes are packed in boxes of 3, how many boxes would you fill?'. Pupils should see that *both* sharing and grouping problems can be solved by doing a division and that division can be thought of in terms of sharing *or* grouping: the division 80 ÷ 4 can be seen as a sharing problem (perhaps solved by halving and then halving again) but 80 ÷ 10 is more likely to be solved as a grouping problem (how many 10s are in 80?).

◊ Ask pupils which division facts they think they know 'by heart' and which they have to think about. Discuss how they work these out.
Some possible methods are
- to divide by 4, halve and halve again
- to divide by 8, halve three times
- to divide by 5, multiply by 2 and divide by 10

E Multiplying by a single-digit number (p 31)

◊ You could begin the discussion by asking pupils how they would calculate '23 × 4'. A mental method (Lee) and three written methods (Sue, Asif and Jim) are shown on the page and these can be discussed.

◊ Encourage mental methods where appropriate.

◊ Any pupils who resort to repeated addition (Sue's method) should be encouraged to move to a more efficient method.

Three digits

◊ With three different digits there are six different multiplications (counting 34 × 6 the same as 6 × 34)

   34 × 6   43 × 6   36 × 4   63 × 4   46 × 3   64 × 3

The largest result is 43 × 6 and the smallest is 46 × 3.

◊ To obtain the largest result, it is clear that the smallest number must go in the units column of the two-digit number (to minimise its effect) giving two possibilities: 43 × 6 or 63 × 4. It must be the first case as 6 × 3 is greater than 4 × 3.
In general, choose the largest digit to be the single-digit multiplier and use the next largest in the tens column.

F Dividing by a single-digit number (p 33)

◊ The methods shown can be compared in class discussion. An advantage of Paula's 'chunking' method is that it can be adapted for dividing by numbers with more than one digit. Of course, Paula and Ted would have found it easier to take away 60 (ten 6s) at a time to start with.

Many pupils will have learned a 'standard' written method. Others may have methods of their own, or develop such methods. They should be encouraged to use the one they feel most confident with, so long as it is not too long-winded.

G What do we do about remainders? (p 35)

◊ Initially each pupil could tackle the three problems on their own. Then they can discuss their solutions in groups.
Each group could make up a problem for the whole class to do.

A Adding and subtracting – mental methods (p 20)

A1 (a) 39 (b) 47 (c) 71 (d) 44
(e) 85 (f) 73 (g) 45 (h) 63
(i) 85 (j) 92

A2 (a) 157 (b) 386 (c) 325 (d) 460
(e) 130 (f) 159 (g) 271 (h) 185
(i) 181 (j) 301

A3 (a) 47 (b) 15 (c) 58 (d) 18
(e) 9

A4 (a) 12 (b) 21 (c) 14 (d) 37
(e) 44 (f) 36 (g) 74 (h) 17
(i) 21 (j) 29

A5 (a) 61 (b) 74 (c) 133 (d) 209
(e) 99

A6 (a) 341 (b) 224 (c) 95 (d) 457
(e) 107

A7 (a) 51 (b) 33 (c) 200 (d) 14
(e) 325 (f) 81 (g) 64 (h) 3
(i) 193 (j) 237

B Adding and subtracting – written methods (p 22)

B1 (a) 195 (b) 319 (c) 372
(d) 680 (e) 684

B2 (a) 432 (b) 332 (c) 282
(d) 781 (e) 1020

B3 (a) 254 (b) 347 (c) 356
(d) 392 (e) 182

B4 (a) 614 (b) 237 (c) 292
(d) 164 (e) 515

B5 (a) 466 (b) 588 (c) 475
(d) 688 (e) 579

B6 (a) 423 (b) 417 (c) 105
(d) 244 (e) 188

B7 (a) 173 (b) 355 (c) 859
(d) 479 (e) 651

B8 (a) 152 (b) 346 (c) 468
(d) 249 (e) 156

B9 (a) £2.91 (b) £2.02
(c) £1.71 (d) £3.19

B10 (a) £1.42 (b) £1.29
(c) £3.25 (d) £1.39

B11 (a) £2.69 (b) £8.67
(c) £3.47 (d) £2.48

B12 (a) £1.60 (b) £4.29 (c) £2.28

B13 £0.35 or 35p

B14 £1.95

B15 £1.57

***B16** Coffee and a cheese ploughman's is the most likely lunch. Other combinations are possible, for example 3 lemonades, a ham sandwich and a tea cake.

C Up to 10 × 10 (p 26)

C1 (a) 20 (b) 35 (c) 24 (d) 21
(e) 28 (f) 27 (g) 80 (h) 40
(i) 42 (j) 64 (k) 56 (l) 54
(m) 49 (n) 36 (o) 72

46 • 3 Whole number calculation

C2 (a)

	5	6	2
7	**35**	**42**	**14**
10	**50**	60	20
1	**5**	**6**	**2**

(b)

	9	3	**5**
8	**72**	**24**	**40**
5	**45**	**15**	25
6	**54**	**18**	**30**

(c)

	3	5	**8**
4	12	**20**	**32**
1	**3**	**5**	**8**
10	**30**	**50**	80

(d)

	9	7	6
9	81	**63**	**54**
8	**72**	**56**	48
7	**63**	**49**	**42**

(e)

	8	5	4
9	**72**	**45**	**36**
6	**48**	30	24
7	**56**	**35**	**28**

(f)

	7	6	9
6	**42**	**36**	**54**
3	**21**	18	**27**
4	**28**	**24**	**36**

Cover up

Using all 18 pieces, the solution is

14	20	27
24	18	36
	21	
32	16	16
	25	12
35	48	
9	42	15	28

The mixed-up table puzzle

Q 9, R 8, S 5, T 1, U 0, V 3, W 7, X 4, Y 6, Z 2

Ⓓ **Division** (p 29)

D1 (a) 4 (b) 5 (c) 8 (d) 5
(e) 4 (f) 7 (g) 9 (h) 3
(i) 5 (j) 4

Links, chains and loops

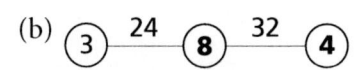

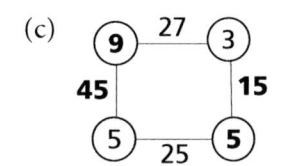

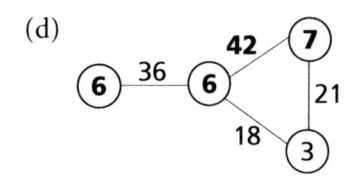

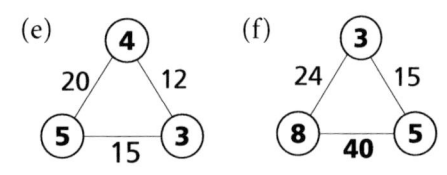

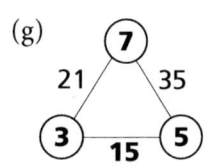

(a) ④ — 28 — ⑦

(b) ③ — 24 — ⑧ — 32 — ④

(c) ⑨ — 27 — ③
    45   15
    ⑤ — 25 — ⑤

(d) ⑥ — 36 — ⑥ — 42 — ⑦
                  18 \ / 21
                     ③

(e) ④
   20 / \ 12
   ⑤ —15— ③

(f) ③
   24 / \ 15
   ⑧ —40— ⑤

(g) ⑦
   21 / \ 35
   ③ —15— ⑤

D2 (a) 8 (b) 6 (c) 6 (d) 9
(e) 9 (f) 8 (g) 7 (h) 7
(i) 7 (j) 9

D3 (a)

	3	6	5	9
8	**24**	**48**	**40**	**72**
2	**6**	**12**	**10**	**18**
4	**12**	**24**	20	**36**
7	**21**	**42**	**35**	**63**

(b)

	7	8	5	9
4	**28**	**32**	**20**	**36**
3	**21**	**24**	15	**27**
6	**42**	**48**	**30**	**54**
2	**14**	**16**	10	**18**

(c)

	9	**6**	**7**	**2**
8	**72**	**48**	**56**	**16**
3	**27**	18	**21**	6
4	**36**	**24**	**28**	8
5	**45**	30	**35**	10

(d)

	6	3	2	4
7	42	**21**	**14**	**28**
9	**54**	**27**	18	**36**
5	**30**	15	10	**20**
8	**48**	24	**16**	**32**

(e)

	3	8	5	**7**
2	**6**	**16**	**10**	**14**
4	**12**	32	20	**28**
9	**27**	**72**	**45**	**63**
6	18	**48**	**30**	**42**

(f)

	4	9	6	5
8	**32**	72	48	40
2	**8**	18	12	10
7	**28**	**63**	**42**	35
3	12	**27**	**18**	15

3 Whole number calculation • 47

(g)

	4	8	9	3
5	20	40	45	15
2	8	16	18	6
6	24	48	54	18
7	28	56	63	21

(h)

	6	3	4	8
9	54	27	36	72
2	12	6	8	16
7	42	21	28	56
5	30	15	20	40

(i)

	4	7	5	8
3	12	21	15	24
9	36	63	45	72
6	24	42	30	48
2	8	14	10	16

D4 (a) 6 (b) 20 (c) 9 (d) 7
(e) 6 (f) 8 (g) 7 (h) 8
(i) 8 (j) 9 (k) 9 (l) 9

D5 (a) 6 remainder 2 (b) 4 remainder 2
(c) 8 remainder 1 (d) 8 remainder 4
(e) 4 remainder 1 (f) 7 remainder 3
(g) 2 remainder 6 (h) 6 remainder 4
(i) 7 remainder 2 (j) 9 remainder 4

E Multiplying by a single-digit number (p 31)

E1 (a) 52 (b) 45 (c) 115 (d) 153

E2 (a) 128 (b) 84 (c) 210 (d) 102
(e) 94 (f) 168 (g) 315 (h) 215

E3 120

E4 212

E5 288

E6 (a) 248 (b) 309 (c) 410
(d) 450 (e) 705 (f) 1278
(g) 921 (h) 864 (i) 774
(j) 1498 (k) 2008 (l) 963
(m) 1224 (n) 2349 (o) 2958

E7 (a) About 108 feet
(b) About 744 metres
(c) About 2667 metres
(d) About 2889 metres
(e) About 1456 feet

F Dividing by a single-digit number (p 33)

F1 £0.58 or 58p

F2 12

F3 28

F4 (a) 16 (b) 23 (c) 17 (d) 39
(e) 14 (f) 19 (g) 15 (h) 12
(i) 12 (j) 13

F5 (a) £1.45 (b) £0.46 or 46p
(c) £1.06 (d) £0.92 or 92p
(e) £0.41 or 41p

F6 (a) 41 (b) 43 (c) 49 (d) 398
(e) 134 (f) 57 (g) 121 (h) 103
(i) 55 (j) 123

F7 (a) £0.45 or 45p (b) £0.05 or 5p

F8 (a) 33 rem 1 (b) 29 rem 2
(c) 31 rem 3 (d) 34 rem 4
(e) 28 rem 1 (f) 78 rem 2
(g) 24 (h) 103 rem 7

F9 19

F10 (a) 81 (b) 128 (c) 26
(d) 14 (e) £446

F11 The pupil's problems in words

G What do we do about remainders? (p 35)

G1 25

G2 12

G3 18

G4 29

G5 17

G6 19

48 • 3 Whole number calculation

G7 The pupil's problem

G8 15

What progress have you made? (p 36)

1. (a) 64 (b) 88 (c) 281
   (d) 16 (e) 38 (f) 89
2. (a) 857 (b) 421 (c) 632
   (d) 514 (e) 316 (f) 177
3. (a) £3.22 (b) £4.91
   (c) £2.87 (d) £1.44
4. (a) 35 (b) 32 (c) 27
   (d) 56 (e) 48 (f) 72
5. (a) 7 (b) 9 (c) 8
   (d) 7 (e) 9 (f) 6
6. (a) 74 (b) 60 (c) 135
7. (a) 87 (b) 365 (c) 564
   (d) 588 (e) 1652 (f) 6318
8. (a) 13 (b) 24 (c) £1.46
   (d) 115 (e) 82 (f) 115
9. 17
10. 32

Practice booklet

Sections A and B (p 7)

1. (a)

+	11	25	22
18	29	43	40
46	57	71	68
44	55	69	66

(b)

+	15	26	17
31	46	57	48
26	41	52	43
40	55	66	57

(c)

+	45	56	63
19	64	75	82
27	72	83	90
34	79	90	97

2. (a) 155 (b) 56 (c) 300 (d) 59

   (e) 47 (f) 33 (g) 26 (h) 19
   (i) 5 (j) 155 (k) 123 (l) 99
3. (a) 953 (b) 413 (c) 978
4. 1804
5. (a) 4 (b) 4 (c) 9
6. (a) 631 (b) 223 (c) 267
   (d) 145 (e) 459 (f) 507
   (g) 259 (h) 45 (i) 381
7. (a) 361 miles (b) 21 miles
8. £4.62
9. (a) 35p or £0.35 (b) 19p or £0.19
10. (a) £8.20 (b) £4.54 (c) £4.13
    (d) £5.41 (e) £9.37 (f) £4.22

Sections C, D, E, F and G (p 8)

1. (a) 12 (b) 30 (c) 27 (d) 56
   (e) 7 (f) 8 (g) 9 (h) 7
   (i) 72 (j) 9 (k) 35 (l) 3
2. (a) £4.32 (b) £2.88 (c) £3.43
   (d) £2.88 (e) £2.08
3. (a) 6 remainder 2 (b) 4 remainder 1
   (c) 8 remainder 6 (d) 7 remainder 3
4. 432
5. £1500
6. (a) 24 (b) 16 (c) 13 (d) 121
   (e) 26 (f) 122 (g) 102 (h) 275
7. £1.14
8. 65
9. 37 with 3 brushes left over
10. 4
11. 7
12. 23
13. Each boy gets more. (Each boy gets 95p and each girl gets 87p.)

3 Whole number calculation • 49

4 Oral questions: time (p 37)   7T/2

The page is for teacher-led oral work on telling the time and carrying out simple calculations with time. It can be used to practise skills that pupils have already acquired and to introduce more demanding mental work on the topic.

◊ Not all pupils are fluent with both of the common time formats ('ten to six' and '5:50') so you should use both in your questioning.

These sample questions are roughly in order of increasing difficulty.

1 What time does watch C show?   4:30
2 Which watch shows five to six?   I
3 Watch A is one hour slow. What is the real time?   4:00
4 Mr Jones takes some trousers in to be dry cleaned. Watch G shows the time. The trousers will be ready in 3 h. When will they be ready?   5:45
5 You look at your watch and it looks like B. How long before it will look like D? $1\frac{1}{4}$ h
6 Claire's friend promised to meet her at two o'clock. G shows the time and her friend still hasn't arrived. How late is she? $\frac{3}{4}$ h
7 You phone your friend at a quarter to nine. Watch D shows the time when you finish the phone call. How long was the phone call? $\frac{1}{2}$ h
8 Mrs Patel parks her car. Watch D shows the time. She pays for 45 minutes' parking. When must she get back to her car?   10:00
9 If the real time is ten past one, how fast or slow is watch F?   20 min slow
10 Sarah needs to catch a train that goes at 7:05. She looks at her watch and H is what she sees. How long before her train goes?   40 min
11 F shows the time when you put a cake in the oven. It takes 50 minutes. When is it ready?   1:40
12 H shows the time when you get on a bus. B shows the time when you get off. How long has your journey taken?   1 h 35 min
13 A TV programme starts at time I and ends at time D. Will a three-hour video tape be long enough to record it?   No, it's 3 h 20 min long
14 Imagine D and E are different watches and you are looking at them at the same time. If D is 20 minutes slow, how fast is E?   1 h 5 min

◊ You can devise questions that relate to the real school day, for example:

1 If A shows the correct time, how long is it until the end of school?
2 Do any of the watches show a time during today's maths lesson?

5 Growing patterns

7T/9, 7C/4

Pupils investigate sequences arising from a variety of contexts.

The emphasis is on finding a rule to *continue* a sequence and explaining why the rule is valid. Work on finding a rule that effectively gives the nth term is covered fully in unit 22 'Work to rule'.

Pupils should begin to realise that just spotting a pattern in the first few numbers in a sequence (for example, 'add 3 to the previous number') is not enough to prove that the sequence will continue in the same way. You have to go back to the context (rose bushes, ponds, earrings, ...) to give a convincing explanation.

p 38	**A** Coming up roses	Investigation leading to a linear sequence Describing how the sequence continues Explaining why the sequence continues like this
p 38	**B** Pond life	Investigation leading to a linear sequence
p 40	**C** Changing shape	Consolidating work on linear sequences
p 41	**D** Earrings	Investigation leading to a simple exponential sequence (powers of 2)
p 41	**E** Staircases	Investigation leading to a Fibonacci sequence

Optional
Coloured counters
Coloured tiles
Multilink cubes
Square dotty paper, triangular dotty paper
Sheets 70, 71 and 72

Practice booklet pages 11 and 12

5 Growing patterns • 51

A Coming up roses (p 38)

This investigation gives rise to a linear sequence.
Pupils describe and explain how the sequence continues.

> Optional: Tiles or counters to represent red and white rose bushes

◊ A gardener is designing a display with red and white rose bushes planted as single rows of red roses, with a row of white roses on each side and a white rose at each end. Some examples are:

🌼 represents a white rose bush.

🌹 represents a red rose bush.

◊ To start with, pupils could think about the designs on the page and then try to draw similar arrangements that use 2, 6, 4 red roses, for example.

Encourage pupils to simplify the diagrams, for example by using coloured circles or the letters R and W.

Pupils count the number of red and white rose bushes needed in each arrangement so far and collect their results together in an ordered table. Discuss the advantages of tabulating in this way.

Alternatively, pupils could consider what the smallest design would look like and draw it. A sequence of designs can then be produced in order: 1 red rose bush, 2 red rose bushes etc. These results can then be tabulated.

◊ Ask pupils to complete the table up to, say, 8 red rose bushes.
Discuss any methods that pupils use to complete the table.
These may include

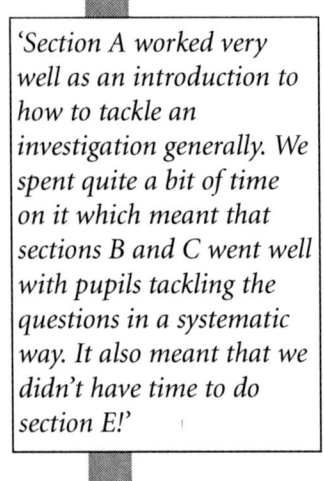

- making or drawing the designs
- finding and using the rule that the number of white rose bushes increases by 2 for every extra red rose bush
- finding and using the rule that the number of white rose bushes is 2 × *the number of red rose bushes* + 2

It is important that each method is considered equally valid.

◊ In discussion, bring out the fact that the number of white rose bushes increases by 2 for every extra red rose bush. Ask the pupils to use the diagrams to explain why this is so.

'Section A worked very well as an introduction to how to tackle an investigation generally. We spent quite a bit of time on it which meant that sections B and C went well with pupils tackling the questions in a systematic way. It also meant that we didn't have time to do section E!'

52 • 5 Growing patterns

You could ask the pupils to consider a range of 'explanations'.
For example, the number of white rose bushes increases by 2 for every extra red rose bush because

- the numbers in the table go up in 2s
- the row of red bushes has 1 white bush at each end (2 in total)
- there are 2 colours of roses
- each red rose bush has 1 white rose bush either side of it (2 in total)

These statements could be put on cards and groups of pupils could consider which is an acceptable explanation.

Many pupils would think that just describing how the sequence continues (as in the first statement above) is a perfectly acceptable 'explanation'. Emphasise they must refer back to the arrangement of bushes to explain why the sequence will continue in the same way.

◊ Now ask pupils to think about a larger number of red bushes, for example 20 red bushes, and to say how many white bushes would be needed for them. Again, pupils are likely to use a variety of methods as before.

If pupils use the rule that the number of white rose bushes is 2 × the number of red rose bushes + 2, ask them to explain why they know their rule works by referring to the arrangement of rose bushes. Emphasise that just because their rule works for a few results it does not follow it will work for all results.

B **Pond life** (p 38)

Pupils investigate another situation that gives rise to a linear sequence. They describe and explain how the sequence continues.

> Optional: Square dotty paper, square tiles

◊ Make sure pupils are clear that the ponds are square.

B5 A variety of methods are possible. If pupils are struggling ask them to look at their answer for B4(d) and think how that could help them. Some pupils will continue to draw ponds at this stage.

B6 In part (b), some pupils may say 'because the numbers in the table go up in 4s'. Emphasise that they must refer back to the arrangement of slabs to explain why they can be sure that the sequence continues in the same way.

B8 Some pupils will solve this problem by counting on in 4s. Encourage the more able to think of a more direct method (for example, multiply by 4 and add 4).

C **Changing shape** (p 40)

This consolidates work on linear sequences.

> Optional:
> Square dotty paper, triangular dotty paper
> Square tiles

C1 In part (b), emphasise that the width of all the ponds for this table is 3 metres. The length of the pond is the other dimension so, for some ponds, the length is shorter than the width.

C5 Some pupils will need help to structure their work.

As an extension, pupils could consider triangular ponds surrounded by triangular slabs.

D **Earrings** (p 41)

Pupils investigate a situation leading to a simple exponential sequence (powers of 2). They find and explain how the sequence continues.

> Optional:
> Red and yellow multilink cubes
> Sheet 70 (for recording results)

◊ Multilink has been found to be a very useful way to 'build' the earrings. It allows easy identification of duplicates and collection of results.

If pupils use multilink, make sure they realise that these two designs count as different.

◊ Ask pupils to find as many different three-bead earrings as they can.

Collect the results for the whole class and discuss how they can be sure that they have found all possible designs for three beads.

The 8 different designs are:

◊ It is important to stress to the more able that 'predict and check' is a good strategy to increase confidence that any patterns or rules found are correct. However, predict and check does not explain *why* any patterns or rules are valid and hence is not a proof that the sequence of numbers will continue in the way they expect.

D3 In part (c), pupils can record their results on sheet 70.

D4 Pupils cannot claim to be sure about the number of five-bead earrings until they have found them all (and shown no more exist) or until they have explained why the numbers in the sequence double each time.

*D6 Most pupils will find it difficult to explain why the number of earrings doubles each time. If so, encourage them to look at their sets of earrings for, say, 3 beads and 4 beads and to consider how they are related.

E Staircases (p 41)

Pupils investigate a situation that gives rise to a Fibonacci sequence.

Optional: Sheets 71 and 72 (for recording results)

◊ Emphasise that the staircase in the diagram has four steps and not five. This can confuse pupils and using sheets 71 and 72 may help.

'Pupils gained a great deal from doing this on a real flight of stairs, as it was easy to see which moves were not allowed.'

◊ Ask pupils to think carefully about ways of recording their results. Pupils who choose to record their results as sequences of 1s and 2s may realise that the problem reduces to that of finding how many different ways a total can be reached by adding 1s and 2s. The importance of being systematic cannot be overemphasised.

E3 Cover up the last two entries in the table and ask pupils to imagine they were trying to predict the number of ways to climb four steps. It's very tempting to predict 4 ways, the sequence 1, 2, 3, … increasing by 1 each time. However, the prediction would be incorrect. This is an opportunity to reinforce the dangers of relying on an apparent number pattern.

B Pond life (p 38)

B1 (a) 8 slabs   (b) 12 slabs   (c) 16 slabs

B2 (a)   (b) 20 slabs

B3 (a)   (b) 24 slabs

B4 (a) The numbers of slabs in the table are
 8, 12, 16, 20, 24, 28.
 (b) 32 slabs   (c)

(d) 44 slabs
 Pupils' methods are likely to involve
 • counting on in 4s or
 • multiplying by 4 and adding 4 or
 • adding 1 and multiplying by 4

B5 11 by 11 metres
 Pupils' methods could involve
 • extending the pattern in the table or
 • working from the fact that a 10 by 10 pond needs 44 slabs or
 • subtracting 4 and dividing by 4 or
 • dividing by 4 and subtracting 1

B6 (a) The number of slabs needed goes up by 4 each time.
 (b) The pupil's explanation: for example, an increase of 1 metre in width means an extra slab for each edge. Since the pond has 4 edges, 4 extra slabs are needed.

B7 276 slabs (272 + 4)

B8 64 slabs
The pupil's method: for example,
15 × 4 + 4 = 64

B9 404 slabs
The pupil's method: for example,
100 × 4 + 4 = 404

*__B10__ 50 by 50 metres
The pupil's method: for example,
(204 − 4) ÷ 4 = 50

*__B11__ The pupil's explanation: for example, an 11 m by 11 m pond needs 48 slabs and a 12 m by 12 m pond needs 52 slabs.

C Changing shape (p 40)

C1 (a) (i) 20 slabs (ii) 14 slabs
(b) The numbers of slabs in the table are 12, 14, 16, 18, 20, 22, 24.
(c) 26 slabs
The pupil's method: for example,
24 + 2 = 26

C2 (a) The number of slabs needed goes up by 2 each time.
(b) The pupil's explanation

C3 50 slabs
The pupil's method: for example,
20 × 2 + 10 = 50

C4 22 m
The pupil's method: for example, take 10 off for the end slabs giving 44 slabs altogether on the 'top' and 'bottom'; divide this by 2 to get the number of slabs along the top, which is the pond length.

C5 The pupil's investigations

D Earrings (p 41)

D1 4 different earrings

D2 The numbers of different earrings in the table are 2, 4, 8.

D3 (a) 16 different earrings
(b) The pupil's method: for example,
8 × 2 = 16 or 8 + 8 = 16
(c) One way to organise the results is to add a yellow bead to each of the three-bead earrings and then add a red bead.

Another way is to look at earrings with 0 red beads, 1 red bead, 2 red beads, …

D4 32 different earrings

D5 The number of earrings doubles each time.

*__D6__ The pupil's explanation

E Staircases (p 41)

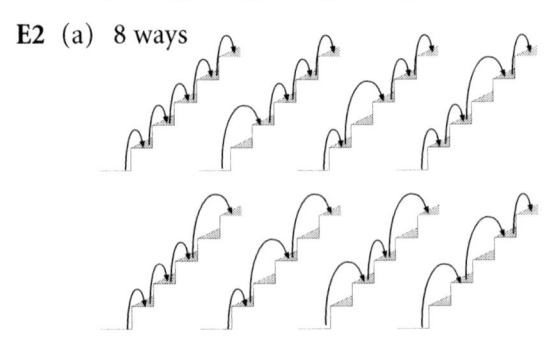

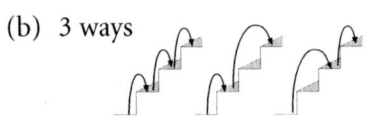

E1 5 ways

E2 (a) 8 ways

(b) 3 ways

56 • 5 Growing patterns

E3 The numbers of different ways in the table are 1, 2, 3, 5, 8.

E4 (a) 13 ways

(b) The pupil's method: for example, add the previous two numbers in the sequence, 5 + 8 = 13.

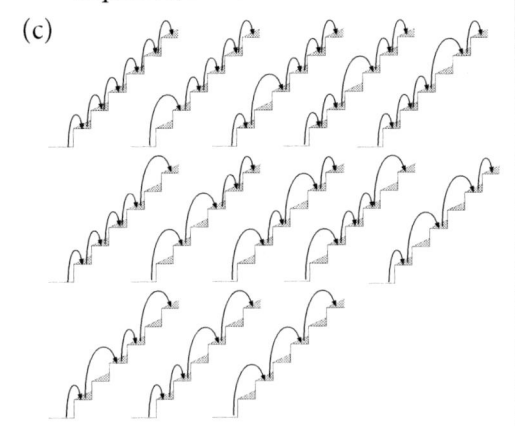

(c)

E5 To find the number of ways to climb a staircase, add the number of ways to climb the previous two staircases.

What progress have you made? (p 42)

1 7 bushes

2 The numbers of yellow rose bushes in the table are 5, 6, 7, 8, 9.

3 (a) 12 yellow bushes

(b) The pupil's method: for example, count in 1s from the result for 5 red bushes.

4 (a) The number of yellow bushes goes up by 1 each time.

(b) The pupil's explanation: for example, an increase of 1 red bush means an extra yellow bush in the bottom row.

5 (a) 104 yellow bushes

(b) The pupil's method: for example, 100 + 4 = 104

Practice booklet

Sections A, B and C (p 11)

1 The pupil's check

2 7 corners

3 The pupil's chain of 4 squares; 13 corners

4 The numbers of corners in the table are **4**, **7**, 10, **13**.

5 (a) 16 corners

(b) The pupil's drawing of a 5-square chain

6 31 corners

7 (a) The number of corners goes up by 3 for each added square.

(b) The corner of a new square that joins the chain has already been counted, so 3 corners are added each time.

8 61 corners

Sections D and E (p 12)

1 (a) 5 ways (b) 8 ways

2 The numbers of different ways in the table are **1**, 2, 3, **5**, **8**.

3 (a) 13 ways

(b) The pupil's method: for example, 5 + 8 = 13

(c) 1, 1, 1, 1, 1, 1   1, 1, 2, 2
     1, 1, 1, 1, 2   1, 2, 1, 2
     1, 1, 1, 2, 1   2, 1, 1, 2
     1, 1, 2, 1, 1   1, 2, 2, 1
     1, 2, 1, 1, 1   2, 1, 2, 1
     2, 1, 1, 1, 1   2, 2, 1, 1
                        2, 2, 2

4 To find the number of different ways, add together the numbers of ways for the previous two values.

5 Growing patterns • 57

6 Test it!

7T/4, 7C/5

Pupils collect measurements to test general statements.

p 43	**A** I don't believe it!	Planning a task Collecting data to test a statement Measuring
p 44	**B** Organising your results	Organising measurements
p 45	**C** Now it's your turn!	Testing a chosen statement

> **Essential**
> Metre sticks, tape measures (enough for at least one item per group)

Ⓐ I don't believe it! (p 43)

> Metre sticks, tape measures (at least one item per group)

◊ Organise pupils into small working groups. (Groups of four work well.)

After you have introduced the statement 'Everyone is six and a half feet tall', the groups can discuss the first set of questions.

It should become clear that

- six and a half feet tall means six and a half 'foot lengths' tall
- the statement can be tested by measuring or simply stepping off each pupil's foot length against their height

You could have a general discussion at this point comparing the groups' plans for testing the statement. Or the groups could move straight into carrying out their plans. Some pupils may need some help with measuring.

Recording of data may be haphazard. This is taken up in section B, which some teachers have preferred to do before A.

Methods used by pupils include:
- making an outline of themselves on a roll of old wallpaper, then cutting out their footprints to test the statement
- Blu-tacking rulers to walls to make it easier to measure heights
- measuring out heights on the tape and then 'stepping off'

In work of this kind it is important to make a plan and to adapt it as necessary. In many cases pupils do not see the need for a plan, preferring instead to 'jump right in'. You may find examples to emphasise this point in the pupils' own work.

Encourage each group to compare findings with others.

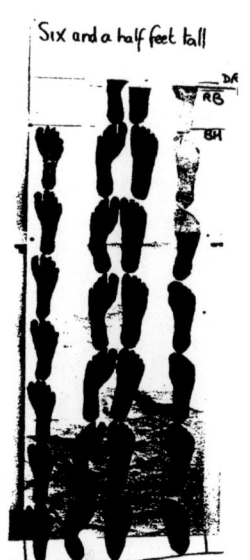

◊ After this discussion it is worth raising the issue of whether the statement 'Everyone is six and a half feet tall' is true for a wider population. Pupils could investigate the statement by measuring younger or older people at home for homework.

> 'Pupils enjoyed this topic. However, there were several teething problems such as which groups they were in and who was responsible for what. When things settled down they produced some excellent work which was good for display. The groups also gave a presentation of their work.'

B Organising your results (p 44)

◊ Some teachers have preferred to do this section before section A.

B1 Pupils should consider the problems of
- mixed units
- writing results in different orders

B2 You may need to help with 'approximately'.

C Now it's your turn! (p 46)

Pupils choose their own general statement to test.

◊ Each group must decide how to measure, for example, the 'length' of a person's head.

◊ A formal write-up is not necessarily expected at this stage. You could ask for a poster from each group or each pupil. (Later in the course there is a more specific focus on writing up results.)

B Organising your results (p 44)

B1 This method can be improved by presenting results in the same order and everyone using the same units.

B2 Ben is correct.

B3 (a) It may mean the height from the chin to the top of the head (when the mouth is closed).

(b) Tim's arm span might be 170 cm.
Gina's height might be 1.44 m.
Sue's height might be 1.59 m.
Sue's foot length might be 26 cm.
Ryan's height might be 1.5 m.
Lara's hand span might be 17 cm.

(c) Tim, Sue, Ryan (if he is 1.5 m tall) and Majid can go on the rides.

(d) Neena's arm span might be about the same length as Ajaz's, 153 cm.

(e) Depends on the pupil's own measurements

(f) About 6.4 to 7.3 times
(The height and head length have to be expressed in the same units before dividing.)

What progress have you made? (p 46)

1 The height is about 9 times the hand length.

7 Angle

7T/7

p 47	**A** Comparing angles	Making the connection between angles and turning Right angles, acute, obtuse and reflex angles
p 50	**B** Measuring angles	
p 52	**C** Drawing angles	
p 53	**D** Angles round a point	
p 54	**E** Calculating angles	Angles round a point, angles on a line, vertically opposite angles

Essential

Sheet 73 (for you: one on white card, one on coloured card)
Sheet 74 (for each pupil: one circle on white card, one on coloured card)
Sheets 76 and 77
Angle measurers
Compasses

Practice booklet pages 13 to 15

A Comparing angles (p 47)

> Use sheet 73 to make a large angle-maker for yourself.
> Each pupil needs a small angle-maker, made from:
> • one circle from sheet 74 on coloured card
> • one circle from sheet 74 on white card
>
> If pupils are to make their own, it is better if they follow a demonstration from you. (You can keep them for other groups.)

◊ You can use the scissors pictures to find out what pupils know already about angle. Do they, for example, think that the size of the scissors affects the size of the angle?

Using the angle-maker

◊ Turn the coloured circle gradually to make each quarter turn like this.

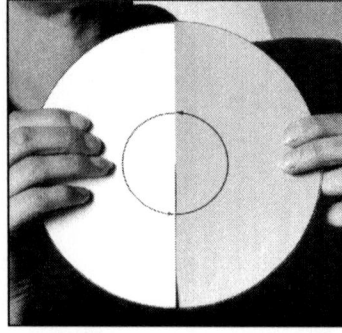

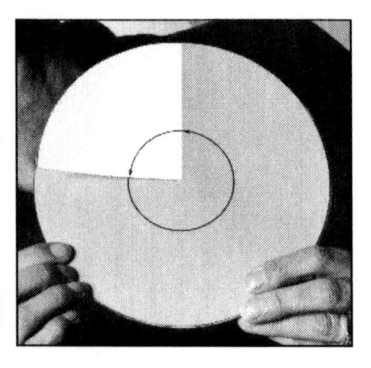

Ask pupils to describe each of the coloured angles in as many ways as they can. Record the words they use and link them together. For example, 90° = right angle = quarter turn.

Set the angle-maker to zero again, and start with the line neither vertical nor horizontal. Ask pupils to tell you to stop turning when you have made

- a right angle
- a half turn
- three quarters of a turn

◊ Introduce the terms acute, obtuse and reflex. You could define them in terms of quarter turn, half turn, etc. or degrees. Make some more angles in various orientations for pupils to describe as acute, obtuse or reflex.

◊ Some pupils see this as a 'large' angle:

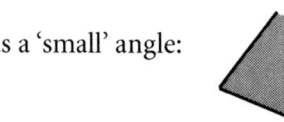

and this as a 'small' angle:

The purpose of the following work is to emphasise that the angle is the amount of turn and has nothing to do with the lengths of the arms or the area between them.

Show an angle on your angle-maker. Ask pupils to make it on their smaller ones. Invite some of them to compare their angle to yours by placing it over yours. Emphasise that the length of the arms doesn't matter.

Make an angle on a large angle-maker and another on a small angle-maker. Ask which is the bigger angle. Include examples where

- the larger angle is on the smaller angle-maker
- both angles are the same
- the larger angle is on the larger angle-maker
- the two angles are held in different orientations

> 'I did do this and it worked very well. I didn't spend too much time on it as almost all of the class knew quite a bit about angles already.'

◊ As a further activity pupils work in pairs, sitting back to back. One makes an angle and describes it as closely as they can. The other makes the angle from the description. Then they compare the angles and see how close they were.

B Measuring angles (p 50)

> Angle measurers (360°), sheets 76, 77

◊ Use an OHP to demonstrate how to use the angle measurer. Show that you can measure starting from either arm. Discuss which scale to use. Link degrees to fractions of a turn.

◊ Draw some angles. Ask pupils to estimate then measure each of them.

Tilting bus

This is to stimulate discussion. The pointer on the body shows how far the body has tilted. The other pointer shows how far the chassis has tilted. The difference is due to the 'give' in the suspension system.

To pass the test, a double decker bus must remain stable when the chassis is tilted to 28° (in practice it can go much further than this). The test has to be done with weights added to simulate a full load of passengers upstairs but no extra weight downstairs.

C Drawing angles (p 52)

> Angle measurers

Some pupils are reluctant to extend lines beyond the point they have marked against the angle measurer.

D Angles round a point (p 53)

> Compasses, angle measurers

E Calculating angles (p 54)

These calculations use angles round a point, angles on a line and vertically opposite angles.

Some pupils can make up their own questions similar to E5 and give them to others to do.

7 Angle • 63

A Comparing angles (p 47)

A1 *a* and *c* are right angles.

A2 (a) Acute (b) Right angle
(c) Acute (d) Obtuse
(e) Obtuse

A3 *a* is a right angle, *b* is acute,
c is acute, *d* is obtuse,
e is acute, *f* is acute.

A4 *a* is acute, *b* is obtuse,
c is acute, *d* is reflex,
e is obtuse, *f* is a right angle.

A5 *a* is acute, *b* is a right angle,
c is reflex, *d* is a right angle,
e is obtuse, *f* is reflex,
g is acute.

B Measuring angles (p 50)

B1 Pupils' answers should be within 1° of the answers given.

Sheet 76
$a = 15°$, acute $b = 90°$, right angle
$c = 50°$, acute $d = 93°$, obtuse
$e = 123°$, obtuse $f = 212°$, reflex
$g = 270°$, reflex $h = 117°$, reflex

Sheet 77
$a = 72°$, acute $b = 58°$, acute
$c = 105°$, obtuse $d = 74°$, acute
$e = 107°$, obtuse $f = 46°$, acute
$g = 113°$, obtuse $h = 54°$, acute
$i = 59°$, acute $j = 60°$, acute

B2 *a* is equal to *e*; *b* is equal to *f*.

B3 (a) $x = 33°, y = 28°, z = 119°$
(b) They add up to 180°.
(c) You should get the same total for any triangle.

B4 (a) $a = 100°, b = 139°, c = 48°, d = 73°$.
(b) They add up to 360°.
(c) You should get the same total for any four-sided shape.

C Drawing angles (p 52)

C1 90°

C2 50°

C3 40°

C4 (a) 80° (b) 360°

D Angles round a point (p 53)

D1 30°

D2 (a) 60° (b) 90° (c) 150°

D3 270°

D4 (a) 150° (b) 210° (c) 270°
(d) 180° (e) 360°

D5 (a) 360° (b) 72°
(c) The pupil's drawing of a regular pentagon
(d) The pupil's check

D6 The angles at the centre are
45° for a regular octagon
40° for a regular nonagon
36° for a regular decagon
30° for a regular dodecagon

E Calculating angles (p 54)

E1 $a = 80°$ $b = 145°$ $c = 60°$ $d = 250°$

E2 $a = 140°$ $b = 35°$ $c = 130°$ $d = 115°$

E3 $a = 39°$ $b = 43°$ $c = 97°$
$d = 234°$ $e = 151°$

E4 $a = 110°$ $b = 70°$ $c = 110°$
$d = 17°$ $e = 163°$ $f = 17°$
$g = 62°$ $h = 25°$ $i = 93°$ $j = 25°$

*E5 $a = 102°$ $b = 36°$ $c = 123°$ $d = 36°$
     $e = 88°$ $f = 129°$ $g = 113°$ $h = 134°$
     $i = 22°$ $j = 125°$ $k = 125°$

What progress have you made? (p 56)

1 a is an acute angle, b is an obtuse angle, c is an acute angle, d is a right angle, e is a reflex angle.

2 a, c, d, b, e

3 $a = 48°$ $b = 112°$ $c = 80°$ $d = 90°$
   $e = 210°$

4 80°

5 $a = 47°$ $b = 110°$ $c = 70°$ $d = 101°$

Practice booklet
Sections A, B and C (p 13)

1 a Acute b Acute c Obtuse
   d Right e Acute

2 e, a, b, d, c

3 The pupil's estimates

4 $a = 32°$ $b = 61°$ $c = 128°$
   $d = 90°$ $e = 14°$

5 The pupil's drawing of an angle of 64°

Section D (p 14)

1 (a) NE   (b) W   (c) W
2 (a) NW   (b) S   (c) NE
3 135°
4 (a) 90°   (b) 270°   (c) 180°
   (d) 135°   (e) 225°   (f) 315°
5 (a) 120°   (b) 150°
6 (a) 210°   (b) 330°

Section E (p 15)

1 $a = 170°$ $b = 60°$ $c = 45°$ $d = 36°$
   $e = 130°$ $f = 105°$ $g = 64°$ $h = 64°$
   $i = 116°$ $j = 60°$ $k = 25°$ $l = 65°$
   $m = 34°$ $n = 80°$ $o = 140°$ $p = 96°$
   $q = 75°$ $r = 72°$ $s = 75°$ $t = 33°$

8 Time

7T/23

p 57	**A** Happiness graphs	Introductory discussion Using a time line
p 57	**B** Time planner	Further use of a time line
p 58	**C** The 24-hour clock	
p 59	**D** How long?	Calculating time intervals
p 60	**E** Round trip	Using timetable information to plan a route
p 61	**F** Timetables	Using timetables

Essential	Optional
Sheet 101	Sheets 98 or 99, and 100 OHP transparencies of 98 or 99, and 101

Practice booklet pages 16 to 18

A Happiness graphs (p 57)

> Optional: Sheet 98 or 99, and a transparency of the sheet used

'Very good. I did my own happiness graph on the board for the day I taught the class.'

◊ Discuss the 'happiness graph'. Happiness is 'measured' on a 0 to 10 scale.

◊ Pupils can draw their own happiness graphs on specially ruled time graph paper (sheet 98 or 99).

Sheet 98
can be used to plot a happiness graph for 9 a.m. to 4 p.m.

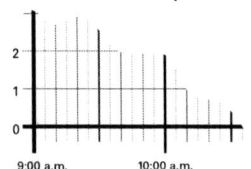

Sheet 99
can be used to plot a happiness graph for 9 a.m. to 11 p.m.

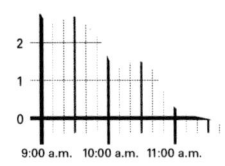

Pupils should label the axes as appropriate, using 12-hour or 24-hour clock times.

◊ Before pupils draw their graphs it may be necessary to establish the times of daily events (e.g. lesson changes, breaks). A transparency of the time graph paper is useful here.

66 • 8 Time

B Time planner (p 57)

> Optional: Sheet 98 or 99, and a transparency of the sheet used

◊ Discuss the diagram on page 57, posing questions such as 'When does assembly end?', 'How long is break?'

◊ Discuss how you could devise a time plan for a whole week by producing a set of bars, one for each day. Pupils can draw diagrams to show their own timetables (including the weekend if they like). A shorter activity is to produce a diagram for the current day only.

If the whole group has the same timetable, each pupil could do one particular school day and then the days could be collected together to make weekly timetables.

Alternatively, pupils could show how they spend a typical Saturday or Sunday.

As before, they can consider the day from 9 a.m. to 4 p.m. (sheet 98) or from 9 a.m. to 11 p.m. (sheet 99). They can use 12-hour or 24-hour clock times.

C The 24-hour clock (p 58)

> Optional: Sheet 100

◊ If you haven't already done so in sections A or B, discuss the 12-hour and 24-hour clocks and how to convert time in one form to the other. Point out that 24-hour clock time can be written as, for example, 16:40 or 1640.

Pupils then order the times A to R. They are reproduced as a set of cards on sheet 100: pupils can cut them out and put them in order.

The correct order is B, Q, A, M, R, O, D, G, N, K, L, I, E, C, P, J, F, H.

Pupils could write all the times as a.m./p.m. and then as 24-hour clock times.

Pupils could also work out the time intervals between adjacent cards when they are in order.

◊ A simple game can be played in groups of three:
 • Shuffle the cards and deal six each.
 • Each player plays a card.
 • The latest time (or earliest, or middle, as agreed) wins the trick. The winner of the trick goes first in the next round.

D How long? (p 59)

D14 You may need to emphasise that each column shows the times for **one** continuous bus journey.

E Round trip (p 60)

The task is to find ways of making a round trip starting and finishing at Midtown, visiting all the other places. Bus times and journey times are given.

> Sheet 101, and a transparency of it

◊ You will probably need to work together through one trip. Emphasise that each time is for a *different* bus. For example, there are four buses from Norsey to Budham at the times shown in the box under Norsey.

You can, if you wish, make further conditions (for example, spend at least 30 minutes at each place).

◊ Pupils can check each other's solutions. The shortest possible trips are these.
- M (10:25) E N B S M   6 hours 5 minutes
- M (10:30) B S E N M   6 hours 30 minutes
- M (9:40) E N B S M   6 hours 50 minutes

However, pupils may not find them in the time available.

F Timetables (p 61)

◊ You may need to emphasise that each column in the timetable shows the times for one continuous bus journey.

You can use the timetable here as the basis for some oral questions similar to the printed questions.

C The 24-hour clock (p 58)

C1 (a) 1430   (b) 1715   (c) 0610
(d) 1225   (e) 2150   (f) 0435
(g) 1825   (h) 2315   (i) 1010
(j) 1940

C2 (a) 7:30 a.m.   (b) 4:00 p.m.
(c) 1:45 p.m.   (d) 1:40 a.m.
(e) 12:50 a.m. (f) 3:15 p.m.
(g) 8:25 a.m.   (h) 12:40 p.m.
(i) 7:55 p.m.   (j) 2:35 p.m.

C3 0205, 6:45 a.m., 11:42 a.m., 1:00 p.m., 1600, 8:35 p.m., 9:40 p.m., 2209

D How long? (p 59)

D1 (a) 9:30 a.m. (b) 10:30 a.m.
   (c) 1 hour

D2 (a) 12:20 p.m. (b) 1:10 p.m.
   (c) 50 minutes

D3 (a) 20 minutes (b) 50 minutes
   (c) 1 hour 10 minutes
   (d) 30 minutes

D4 (a) 1 hour 40 minutes
   (b) 2 hours 30 minutes
   (c) 2 hours 50 minutes
   (d) 1 hour 50 minutes

D5 (a) 30 minutes (b) 15 minutes
   (c) 1 hour 15 minutes
   (d) 1 hour 40 minutes

D6 1 hour 40 minutes

D7 35 minutes

D8 15 minutes

D9 35 minutes

D10 10 minutes

D11 1:45 p.m.

D12 3:25 p.m.

D13 35 minutes

D14 (a) 1 hour 15 minutes
   (b) 1 hour 40 minutes

D15 25 minutes

F Timetables (p 61)

F1 2030 or 8:30 p.m.

F2 25 minutes

F3 4 minutes

F4 22 minutes

F5 13 minutes

F6 The 1934 train

F7 The 2034 train

F8 Four trains

F9 1956 or 7:56 p.m.

F10 1832 or 6:32 p.m.

What progress have you made? (p 62)

1 (a) 1500 (b) 1815
   (c) 1335 (d) 0905

2 (a) 5:00 p.m. (b) 2:20 p.m.
   (c) 6:45 a.m. (d) 10:10 p.m.

3 3:30 a.m., 1135, 2:05 p.m., 2340

4 (a) 45 minutes
   (b) 1 hour 20 minutes
   (c) 13 hours 20 minutes
   (d) 6 hours 25 minutes

5 20 minutes

6 22 minutes

7 1430 or 2:30 p.m.

Practice booklet

Sections C and D (p 16)

1 (a) 1320 (b) 0420 (c) 1150
   (d) 1725 (e) 0900 (f) 2330
   (g) 2105 (h) 1030

2 (a) 6:50 a.m. (b) 1:45 p.m.
   (c) 11:10 a.m. (d) 3:10 p.m.
   (e) 4:20 p.m. (f) 6:35 p.m.
   (g) 8:00 p.m. (h) 10:15 p.m.

3 (a) 15 minutes (b) 25 minutes
   (c) 55 minutes (d) 20 minutes
   (e) 1 hour 25 minutes or 85 minutes
   (f) 1 hour 30 minutes or 90 minutes
   (g) 1 hour 30 minutes or 90 minutes
   (h) 6 hours 20 minutes

4 (a) 20 minutes
   (b) 1 hour 15 minutes or 75 minutes
   (c) 20 minutes
   (d) 50 minutes
   (e) 2 hours 45 minutes

5 (a) 45 minutes
   (b) 1 hour 20 minutes or 80 minutes
   (c) 1 hour 20 minutes or 80 minutes
   (d) 45 minutes
   (e) 2 hours 15 minutes
   (f) 1 hour 20 minutes

6 (a) 45 minutes   (b) 10 minutes
   (c) 1 hour 30 minutes or 90 minutes

Section F (p 17)

1 (a) 30 minutes   (b) 25 minutes
   (c) 20 minutes   (d) 45 minutes

2 (a) (i) 0629 or 6:29 a.m.
       (ii) 14 minutes
   (b) (i) 1445 or 2:45 p.m.
       (ii) 25 minutes
   (c) 1413 or 2:13 p.m.
   (d) 1605 or 4:05 p.m.
   (e) 2 hours 1 minute or 121 minutes
   (f) (i) 55 minutes
       (ii) 1 hour 15 minutes or 75 minutes
   (g) Hereford   2345
       Ledbury   0001
       Colwall   0008
       Great Malvern   0027
       Malvern Link   0030
       Foregate St   0040
       Shrub Hill   0043

9 Place value and rounding 7T/14

T	p 63	**A** Place value
T	p 64	**B** Ordering
T	p 66	**C** Multiplying by 10, 100, 1000
T	p 67	**D** Dividing by 10, 100, 1000
T	p 68	**E** Numbers in the news — Reasons for rounding
	p 70	**F** Rounding to the nearest ten
T	p 71	**G** Rounding to the nearest hundred
T	p 72	**H** Further rounding

Essential
Sheets 19, 26, 27

Optional
Counters
Large cards with the digits 0 to 9 on them

Practice booklet pages 19 to 21

A Place value (p 63)

> Sheet 26 (Bingo cards), sheet 27
> Optional: counters

◊ To play 'Place value bingo', each pupil chooses a card or couple of cards. The caller reads out numbers in random order from the checklist at the top of sheet 26, ticking them off as they go. Pupils cover with counters the numbers they hear (or they can cross them off in pencil if you think errors are likely and want to be able to trace them). The first pupil to cover all the numbers on a card and shout 'bingo' is the winner.

Pupils can make their own bingo card by choosing ten different numbers from a set that you provide for them.

◊ Include some numbers with more than four digits in your discussion.

B Ordering (p 64)

> Two sets of 0–9 cards per pair of pupils (made from sheet 19)

◊ You can make the rules of the 'Bigger wins' game clear by playing a demonstration game on the board with a pupil.

After playing for a while, most pupils realise they should put high numbers in the hundreds box and low numbers in the units box (and vice versa when putting cards in their opponent's boxes in the 'Nasty game'). Ask them to explain why.

C Multiplying by 10, 100, 1000 (p 66)

This is likely to need careful introduction. The well-known rule 'to multiply by 10 you add a 0' has limitations (it will not work for decimals). It is better to think in terms of place value.

> Optional: Large (e.g. A4) cards with digits on them (0 to 9) and some extra 0s

◊ If there is enough room, you can use pupils as digits. Rule four pupil-width columns on the board (units, tens, hundreds and thousands). Give some pupils a large card each with a digit on it (0 to 9). Ask them to make a number, for example 235, against the board. Then give instructions for the group as a whole to respond to, for example:

- 'Add 40' (the pupil with 7 replaces the pupil with 3)
- 'Multiply by 10' (pupils move one place to left; pupil with 0 stands in units place)
- 'Multiply by 100' (pupils move two places to left; pupils with 0 stand in units and tens places)

Encourage pupils to say the numbers aloud to help consolidate work from the previous sections.

◊ Many pupils will hold very strongly to the 'add a zero' strategy.
It may be beneficial for some pupils to consider a few decimal examples (e.g. $1.5 \times 10 \neq 1.50$).

D Dividing by 10, 100, 1000 (p 67)

The well-known 'rule' that to divide by 10 you take off a zero has clear limitations. Not all numbers end in a zero and it will not be appropriate for some that do (e.g. 3.90).

> Optional: Large (e.g. A4) cards with the digits 0 to 9 on them

◊ The activities described in the previous section for multiplying by 10, 100 and 1000 can be adapted to division.

E Numbers in the news (p 68)

◊ Each news extract has an exact number in the text but a rough approximation in the headline. Ask why this is done.
You could give some exact numbers and ask for suggestions for rough approximations. The short exercise includes this and also working the other way from a headline to a possible exact figure.

E3 At this stage, you could look at a number of different answers to each part of the question and discuss what the largest or smallest figure in the story might be for the headline still to apply.

For example, in part (a) if the cyclist's actual trip were 3690 miles, the headline would be more likely to say '3700'. Where would the dividing line be between headlines of 3600 and 3700?

◊ A further activity is for pupils themselves to look for headlines where figures have been rounded.

F Rounding to the nearest ten (p 70)

G Rounding to the nearest hundred (p 71)

◊ By the end of the discussion, pupils should understand that the value of the digit in the tens column tells you whether to round up or down to the nearest hundred.

H Further rounding (p 72)

◊ By the end of the discussion, pupils should understand that the value of the digit in the hundreds column tells you whether to round up or down to the nearest thousand.

A Place value (p 63)

A1 (a) 5020 (b) 1402

A2 (a) Nine thousand, eight hundred and forty-one
 (b) Eight thousand and five
 (c) Nine thousand and thirteen
 (d) Twelve thousand, seven hundred and forty

A3 (a) 5 hundreds or 500
 (b) 7 thousands or 7000
 (c) 8 tens or 80
 (d) 8 hundreds or 800
 (e) 6 tens or 60
 (f) 90 thousands, 9 ten thousands or 90 000

A4 (a) 4583 (b) 4673 (c) 5573
 (d) 6197 (e) 47 097 (f) 46 107

A5 (a) 5723 (b) 2385 (c) 1304
 (d) 5062 (e) 8395 (f) 4002

A6 (a) 7090 (b) 1990 (c) 2900

A7 (a) Ten
 (b) A 210, B 350, C 490, D 560

A8 (a) Hundred
 (b) A 4200, B 5100, C 6500, D 7900

A9 (a) The pupil's arrows on the number line
 (b) The pupil's arrows on the number line
 (c) A 4250, B 4330, C 4480, D 5300, E 7100, F 7900, G 880, H 1030, I 1110, J 1220

B Ordering (p 64)

B1 279, 283, 297, 317, 401

B2 (a) 6431 (b) 1346

B3 8205, 8250, 8502, 8520

B4 (a) 842, 886, 954, 1007, 1026
 (b) 857, 949, 1009, 1030, 1153
 (c) 798, 1260, 1327, 1402, 2003
 (d) 2907, 2913, 3009, 3015, 3104

B5 3377, 3737, 3773, 7337, 7373, 7733

B6 Scafell Pike, Scafell, Helvellyn, Broad Crag, Skiddaw, Ill Crag, Helvellyn Low Man

B7 Gulf of Mexico, Bering Sea, Mediterranean Sea, Arctic Ocean, Malay Sea, Caribbean Sea, Indian Ocean, Atlantic Ocean, Pacific Ocean

C Multiplying by 10, 100, 1000 (p 66)

C1 (a) 260 (b) 2600 (c) 4800
 (d) 32 400 (e) 6600 (f) 6000
 (g) 3600 (h) 84 000 (i) 2400
 (j) 661 000

C2 (a) 500 (b) 4000 (c) 6010
 (d) 50 000 (e) 80 600 (f) 4200
 (g) 5000 (h) 50 000 (i) 65 000
 (j) 10 000

C3 (a) Four thousand and sixty
 (b) Twenty thousand, five hundred

C4 (a) 640 000, 3000, 30 000, 640 000 → ENSE → SEEN
 (b) 300, 64 000, 30 000, 30 000 → APSS → PASS (or SPAS, ASPS, SAPS)
 (c) 6400, 3000, 300 000, 64 000, 30 → INTPO → POINT
 (d) 300 000, 640 000, 64 000, 640, 300, 300 000 → TEPRAT → PATTER

D Dividing by 10, 100, 1000 (p 67)

D1 (a) 45 (b) 320 (c) 4570
(d) 13 (e) 3600 (f) 830
(g) 8000 (h) 102 (i) 38
(j) 620 (k) 200 (l) 59
(m) 290 (n) 300 (o) 23
(p) 1300

D2 (a) 600 ÷ 10 = 60
30 ÷ 10 = 3
450 ÷ 10 = 45
60 ÷ 10 = 6
780 ÷ 10 = 78

(b) 600 ÷ 100 = 6
3000 ÷ 100 = 30
78 000 ÷ 100 = 780
60 000 ÷ 100 = 600
45 000 ÷ 100 = 450

(c) 3000 ÷ 1000 = 3
78 000 ÷ 1000 = 78
60 000 ÷ 1000 = 60
45 000 ÷ 1000 = 45

D3 (a) 19 (b) 1900
(c) 82 (d) 16 000
(e) 600 (f) 3620
(g) 36 200 000 (h) 20 000

D4 (a) 100 (b) 1000 (c) 10
(d) 60 (e) 7000 (f) 876 000

E Numbers in the news (p 68)

E1 The rounded number is easier for the reader to 'take in', makes a greater impact and is easier to remember.

E2 The pupil's headlines including
(a) 37 000 or 40 000
(b) 25 000 (c) 2000
(d) 370 000 or nearly 400 000

E3 (a)–(d) The pupil's sentences

F Rounding to the nearest ten (p 70)

F1 (a) 50 (b) 30 (c) 80
(d) 60 (e) 60

F2 (a) 140 (b) 270 (c) 440
(d) 860 (e) 710

F3 (a) 100 (b) 500 (c) 300
(d) 110 (e) 390

F4 (a) 70 (b) 40 (c) 90
(d) 100 (e) 20

F5 (a) 180 (b) 290 (c) 500
(d) 610 (e) 360

F6 (a) 1380 (b) 1390 (c) 1400
(d) 1400 (e) 1410

F7 (a) 4620 (b) 6390 (c) 5000
(d) 2180 (e) 3870 (f) 2200
(g) 3010 (h) 4740 (i) 2450
(j) 2000

G Rounding to the nearest hundred (p 71)

G1 (a) 6300 (b) 5900 (c) 6000
(d) 6000 (e) 6100

G2 (a) 600 (b) 800 (c) 100
(d) 100 (e) 800

G3 (a) 1900 (b) 1900 (c) 1800
(d) 52 400 (e) 62 100

G4 (a) 700 (b) 1400 (c) 1900
(d) 2500 (e) 25 000

G5 Pierre St-Martin 4400
Jean Bernard 4300
Cellagua 3200
Corchia 3100
Kacherlschacht 3000
Holloch 2700
Snieznej 2600

9 *Place value and rounding* • 75

G6 (a) 300 (b) 900 (c) 900
 (d) 1200 (e) 2400 (f) 9100
 (g) 12 700 (h) 32 700 (i) 54 800
 (j) 3000

H Further rounding (p 72)

H1 (a) 6000 (b) 12 000 (c) 19 000
 (d) 20 000 (e) 47 000 (f) 5000
 (g) 75 000 (h) 286 000 (i) 842 000
 (j) 740 000

H2 (a) Manchester Utd, Liverpool, Arsenal, Newcastle Utd, Aston Villa, Chelsea
 (b) Newcastle Utd and Aston Villa appear equal (both 36 000).

H3

Division	Attendance
Premier	10 805 000
1	6 932 000
2	3 195 000
3	1 852 000

H4 80

H5 1800 miles

H6 4800 metres

H7 91 000 miles

What progress have you made? (p 73)

1 (a) 6008 (b) 9013

2 (a) 6 tens or 60 (b) 4095

3 A 2600, B 3100, C 4400

4 2031, 2130, 2310, 3002, 3020

5 (a) 6700 (b) 20 000 (c) 34
 (d) 900

6 (a) 70 (b) 570 (c) 2480
 (d) 2700

7 (a) 900 (b) 2100 (c) 25 100
 (d) 2000

8 (a) 2000 (b) 7000 (c) 17 000

Practice booklet

Sections A and B (p 19)

1 (a) 4040 (b) 5004

2 (a) 70 or 7 tens (b) 70 or 7 tens
 (c) 7000 or 7 thousands
 (d) 700 or 7 hundreds

3 (a) 3467 (b) 3447 (c) 3557
 (d) 2457 (e) 7890 (f) 8000

4 (a) 4137 (b) 6305 (c) 6003

5 (a) 5100 (b) 5800 (c) 6900
 (d) 7700 (e) 8600

6 (a) 3840 (b) 3910 (c) 4030
 (d) 4100 (e) 4150

7 (a) 3492, 3509, 3516, 3915, 4065
 (b) 1079, 1097, 1099, 1107, 1119

8 4055, 4505, 4550, 5045, 5054, 5405, 5450, 5504, 5540

Sections C and D (p 20)

1 (a) 370 (b) 70 800 (c) 16 000
 (d) 3000 (e) 20 800 (f) 50 000
 (g) 400 000 (h) 10 000

2 (a) 69 × 10 = **690**
 (b) 37 × 100 = **3700**
 (c) 10 × 45 = **450**
 (d) 10 × **70** = 700
 (e) **65** × 1000 = 65 000
 (f) **50** × 100 = 5000
 (g) 320 × **10** = 3200
 (h) **100** × 90 = 9000
 (i) 70 × **100** = 7000

3 3 × 10 = 30   3 × 1000 = 3000
 30 × 100 = 3000   42 × 10 = 420
 42 × 100 = 4200   420 × 10 = 4200

4 £70

76 • *9 Place value and rounding*

5 (a) 22 (b) 40 (c) 8
  (d) 50 (e) 38 (f) 6500

6 400 ÷ 100 = 4   30 000 ÷ 100 = 300
  300 ÷ 100 = 3   4000 ÷ 100 = 40
  6500 ÷ 100 = 65   2500 ÷ 100 = 25
  3000 ÷ 100 = 30

7 (a) 12 (b) 12 000 (c) 2000
  (d) 340 000 (e) 42 (f) 206

Sections E, F, G and H (p 21)

1 The pupil's headline such as '14 000 foxes killed by hunting in 2000'

2 (a) 720 (b) 390 (c) 550
  (d) 7070 (e) 450 (f) 3080
  (g) 400 (h) 2090

3
	Calais	Santander
Barcelona	900	400
Alicante	1200	500
Marbella	1400	600
Lisbon	1300	600
Faro	1400	700

4 2380, 2231, 2394, 2376, 2351

5 (a) 9000 (b) 1000 (c) 33 000
  (d) 25 000 (e) 13 000 (f) 873 000
  (g) 436 000 (h) 300 000 (i) 85 000
  (j) 405 000

6 5900 metres

7
Cruise ship	Size in gross tons
Arcadia	64 000
Dawn Princess	77 000
Royal Princess	45 000
Minerva	13 000
Oriana	70 000
Island Princess	20 000
Grand Princess	109 000

8 120 miles

9 7000 kilometres

Review 1 (p 74)

Essential Angle measurer

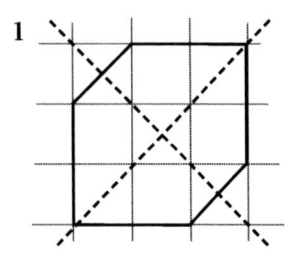

1

2 (a) **35** + **46** = 81 or **45** + **36** = 81
 (b) **63** − **45** = 18 or **54** − **36** = 18
 (c) **34** × **6** = 204   (d) **465** ÷ **3** = 155

3 (a) 16
 (b) The numbers in the table are

1	2	3	4	5	6
8	10	12	14	16	18

 (c) 26   (d) 12
 (e) (i) The number of yellows goes up in 2s.
   (ii) The pupil's explanation
 (f) (i) 46
   (ii) The pupil's working such as 20 × 2 + 6 = 46

4 (a) The pupil's drawing (b) 85° (c) 2

5 1 hour 50 minutes

6 (a) 1320   (b) 0545   (c) 2310

7 (a) 100   (b) 620   (c) 1000

8 19

9 850 metres

10 $a = 145°$ $b = 15°$ $c = 154°$ $d = 26°$
   $e = 122°$ $f = 132°$ $g = 145°$ $h = 145°$
   $i = 35°$ $j = 251°$

11 (a) 8:40 a.m.   (b) 3:25 p.m.
   (c) 10:05 a.m.

12 (a)

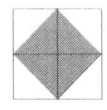

 (b) Excluding rotations there are six different solutions: the pupil's four from

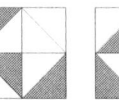

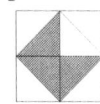

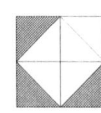

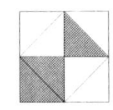

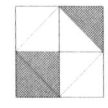

13 9000 metres

14 (a) 25 min   (b) 37 min   (c) 7 min

Mixed questions 1 (Practice booklet p 22)

1 A, C

2 (a) 15 kg   (b) 111 kg
  (c) Maddie 60 kg, Claire 50 kg

3 55 minutes

4 (a) 643   (b) 215   (c) 178

5 (a) A drawing of the plant with 8 flowers
  (b) The completed table with 1, 2, 4, 8 in the bottom row
  (c) It doubles each year.
  (d) The pupil's explanation such as: 'Each flower is replaced by 2 flowers so the number of flowers doubles.'

6 10:40 p.m.

7 27 tables with 2 legs left over

8 6400 metres

9 $a = 118°$ $b = 137°$ $c = 43°$
  $d = 137°$ $e = 249°$

10 288

11 (a) 230 000   (b) 640

12 8000 kg

78 • Review 1

10 Using fractions

7T/11, 7T/39, 7C/8

p 76	**A** Fractions everywhere	Informal ideas of fractions
p 77	**B** Halves, quarters and what else?	Identifying fractions
p 79	**C** Imagine them	Less obvious fractions
p 80	**D** Eighths	Equivalence, simple addition and subtraction
p 81	**E** Fractions of numbers	
p 83	**F** Pie charts	
p 85	**G** Puzzles and problems	

Essential

Sheet 38
Squared paper or square dotty paper
Triangular dotty paper

Practice booklet pages 24 to 27

A Fractions everywhere (p 76)

◊ Before starting, you could ask pupils what they already know about fractions. They can discuss this in pairs before you collect ideas together.

The pictures on this page are intended to stimulate discussion about the meaning of simple fractions used in everyday life. Ask pupils to be as precise as possible when they explain what each fraction means.

B Halves, quarters and what else? (p 77)

Sheet 38, squared paper or square dotty paper, triangular dotty paper

◊ Pupils could work in pairs first, with a class discussion to follow. Focus on pupils' explanations of their answers.

C introduces the idea that a fraction may be made up of separated parts. Pupils who say P is $\frac{1}{2}$ coloured and J and K are $\frac{1}{4}$ coloured are counting parts regardless of whether they are equal.

The shapes that are half coloured are A, B, C, E and F.

Those that are a quarter coloured are D, H, I and, less obviously, P.

The more obvious other fractions are G $\frac{1}{3}$, K $\frac{1}{6}$, L $\frac{3}{4}$, M $\frac{1}{16}$ and N $\frac{1}{6}$.
J $\frac{1}{8}$ and O $\frac{1}{8}$ are much less obvious.

C Imagine them (p 79)

◊ After this section you could discuss shapes such as J, K and P in section B with pupils who found them difficult first time round.

D Eighths (p 80)

Equivalent fractions, which should have arisen naturally in earlier sections, are now dealt with explicitly and used for simple addition and subtraction. Unit 25 'Equivalent fractions and ratio' takes these ideas further.

E Fractions of numbers (p 81)

F Pie charts (p 83)

The pie charts all show simple fractions.

F11 The amounts are not all whole numbers of pounds because $\frac{1}{12}$ of £30 is £2.50.

F12 Again, the amounts are not all whole numbers.

G Puzzles and problems (p 85)

G2 Completing the second and third columns requires 'thinking backwards': $\frac{2}{3}$ of a number is 40, so what is the number? This type of thinking is also needed in G4 (unless it is solved by trial and error).

B Halves, quarters and what else? (p 77)

B1 (a) $\frac{5}{6}$ (b) $\frac{3}{5}$ (c) $\frac{3}{8}$ (d) $\frac{5}{12}$

B2 The pupil's shading of the shapes

B3 The pupil's attempt; an accurate mark would be 2 cm from A.

B4 No, a quarter of the square is green

B5 The pupil's shape, showing $\frac{2}{5}$

B6 The pupil's shape, showing $\frac{2}{7}$

B7 Yes $\frac{3}{9}$ is yellow, and this is the same as $\frac{1}{3}$.

B8 The pupil's shaded shapes

C Imagine them (p 79)

C1 (a) $\frac{1}{8}$ (b) $\frac{3}{8}$ (c) $\frac{3}{8}$ (d) $\frac{5}{8}$

C2 (a) $\frac{1}{6}$ (b) $\frac{2}{6}$ or $\frac{1}{3}$ (c) $\frac{4}{6}$ or $\frac{2}{3}$

C3 (a) $\frac{1}{9}$ (b) $\frac{4}{9}$ (c) $\frac{6}{9}$ or $\frac{2}{3}$
    (d) $\frac{1}{16}$ (e) $\frac{1}{8}$ (f) $\frac{2}{12}$ or $\frac{1}{6}$

C4 Thailand $\frac{2}{6}$ or $\frac{1}{3}$, Czech Republic $\frac{3}{8}$

D Eighths (p 80)

D1 (a) $\frac{4}{8}$ or $\frac{2}{4}$ or $\frac{1}{2}$ (b) $\frac{2}{8}$ or $\frac{1}{4}$
(c) $\frac{6}{8}$ or $\frac{3}{4}$ (d) $\frac{4}{8}$ or $\frac{2}{4}$ or $\frac{1}{2}$
(e) $\frac{4}{8}$ or $\frac{2}{4}$ or $\frac{1}{2}$ (f) $\frac{6}{8}$ or $\frac{3}{4}$

D2 (a) $\frac{6}{8}$ or $\frac{3}{4}$ (b) $\frac{6}{8}$ or $\frac{3}{4}$
(c) $\frac{6}{8}$ or $\frac{3}{4}$ (d) $\frac{4}{8}$ or $\frac{2}{4}$ or $\frac{1}{2}$
(e) $\frac{3}{8}$ (f) $\frac{3}{8}$
(g) $\frac{4}{8}$ or $\frac{2}{4}$ or $\frac{1}{2}$ (h) $\frac{5}{8}$
(i) $\frac{2}{8}$ or $\frac{1}{4}$ (j) $\frac{7}{8}$
(k) $\frac{5}{8}$ (l) $\frac{7}{8}$

E Fractions of numbers (p 81)

E1 (a) 5 (b) 4 (c) 6 (d) 10
(e) 7

E2 (a) 2 (b) 5 (c) 4 (d) 1
(e) 10 (f) 6 (g) 9 (h) 7
(i) 11 (j) 8

E3 (a) 3 (b) 6 (c) 1 (d) 4
(e) 8 (f) 9 (g) 7 (h) 12
(i) 5 (j) 10

E4 (a) To work out $\frac{1}{5}$ of a number, you divide it by **5**.
(b) (i) 3 (ii) 2 (iii) 6

E5 (a) 3 (b) 7 (c) 7
(d) 9 (e) 5

E6 (a) 25 kilograms (b) 8 centimetres
(c) 20 grams (d) 9 litres
(e) 7 kilograms (f) 12 hectares

E7 (a) 9 (b) 18 (c) 12
(d) 3 (e) 30

E8 (a) D (b) E (c) B
(d) F (e) A (f) C

E9 (a) 4 (b) 8 (c) 24 (d) 6
(e) 10 (f) 32 (g) 15 (h) 6

E10 12

E11 $\frac{5}{8}$ of 160 g (100 g) is more than $\frac{3}{4}$ of 120 g (90 g).

E12 $\frac{3}{8}$ of 240 g (90 g) is more than $\frac{2}{5}$ of 200 g (80 g).

E13 (a) 79 (b) 158

E14 (a) 576 (b) 314 (c) 204 (d) 270
(e) 110 (f) 540 (g) 225 (h) 616

E15 (a) 39 litres (b) 120 cm (c) 175 g

F Pie charts (p 83)

F1 (a) $\frac{2}{6}$ or $\frac{1}{3}$ (b) $\frac{1}{6}$

F2 (a) £4 (b) £12 (c) £8

F3 (a) $\frac{1}{8}$ (b) $\frac{3}{8}$ (c) $\frac{2}{8}$ or $\frac{1}{4}$

F4 (a) £2 (b) £6 (c) £4

F5 (a) $\frac{1}{10}$ (b) $\frac{3}{10}$ (c) $\frac{4}{10}$ or $\frac{2}{5}$

F6 (a) 3 (b) 9 (c) 12

F7 (a) $\frac{1}{8}$ (b) $\frac{5}{8}$ (c) $\frac{2}{8}$ or $\frac{1}{4}$

F8 (a) 20 (b) 8

F9 18

F10 14

F11 (a) £10 (b) £12.50 (c) £7.50

F12 (a) 62.5 ha (b) 312.5 ha (c) 125 ha

G Puzzles and problems (p 85)

G1 (a) $\frac{1}{3}$ of 24 (b) $\frac{3}{4}$ of 40 (c) $\frac{3}{4}$ of 36
(d) $\frac{1}{3}$ of 36 and $\frac{2}{5}$ of 30
(e) $\frac{1}{3}$ of 30 and $\frac{2}{5}$ of 25
(f) $\frac{2}{3}$ of 24 and $\frac{2}{5}$ of 40

G2

	12	36	60
$\frac{1}{4}$ of	3	9	15
$\frac{1}{6}$ of	2	6	10
$\frac{2}{3}$ of	8	24	40

G3 He has one gold piece left.

G4 32 trees

10 Using fractions • 81

What progress have you made? (p 86)

1. (a) $\frac{2}{5}$ (b) $\frac{1}{8}$ (c) $\frac{4}{9}$
   (d) $\frac{5}{6}$ (e) $\frac{1}{3}$ (f) $\frac{2}{3}$
2. (a) 5 (b) 12 (c) 12
   (d) 125 (e) 294 (f) 105
3. (a) $\frac{2}{6}$ or $\frac{1}{3}$ (b) $\frac{4}{6}$ or $\frac{2}{3}$

Practice booklet

Sections B and C (p 24)

1. (a) $\frac{1}{6}$ (b) $\frac{3}{8}$ (c) $\frac{2}{8}$ or $\frac{1}{4}$
   (d) $\frac{2}{6}$ or $\frac{1}{3}$ (e) $\frac{3}{6}$ or $\frac{1}{2}$ (f) $\frac{2}{5}$
   (g) $\frac{3}{8}$
2. (a) $\frac{2}{8}$ or $\frac{1}{4}$ (b) $\frac{7}{8}$ (c) $\frac{3}{4}$
   (d) $\frac{3}{5}$ (e) $\frac{7}{16}$ (f) $\frac{1}{2}$
   (g) $\frac{3}{8}$ (h) $\frac{6}{8}$ or $\frac{3}{4}$ (i) $\frac{4}{6}$ or $\frac{2}{3}$

Sections D and E (p 25)

1. (a) Two answers from $\frac{6}{12}, \frac{3}{6}, \frac{2}{4},$ or $\frac{1}{2}$
   (b) $\frac{9}{12}, \frac{3}{4}$ (c) $\frac{4}{12}, \frac{1}{3}$
   (d) $\frac{7}{12}$ (e) $\frac{10}{12}, \frac{5}{6}$
2. (a) 19, 20, 5, 13 → STEM
   (b) 13, 1, 14, 25 → MANY
   (c) 3, 8, 9, 12, 4 → CHILD
   (d) 6, 9, 18, 5, 4 → FIRED
   (e) 12, 15, 12, 12, 25 → LOLLY
   (f) 14, 9, 7, 8, 20 → NIGHT

3. (a) 6 (b) 50 (c) 12
   (d) 6 (e) 10 (f) 32
4. 3
5. (a) $\frac{3}{8}$ of a 240 g bar
   (b) $\frac{3}{5}$ of a packet with 60 sweets
   (c) $\frac{5}{7}$ of a £140 prize
   (d) $\frac{3}{4}$ of a 840 km² of land
6. (a) 369 (b) 430 (c) 469 (d) 87
   (e) 365 (f) 710 (g) 276 (h) 252

Section F (p 26)

1. (a) (i) $\frac{5}{8}$ (ii) $\frac{1}{4}$ or $\frac{2}{8}$
   (b) (i) 5 (ii) 25
2. (a) 4 (b) 8 (c) 16
3. (a) 4 kg (b) 8 kg (c) 20 kg

Section G (p 27)

1. (a) 192 (b) 144 (c) 40
   (d) 7 (e) 1
*2. (a) 10 (b) 5 (c) 4
   (d) 1, which can be returned to the neighbour!
   $\frac{1}{2} + \frac{1}{4} + \frac{1}{5} = \frac{19}{20}$, not a whole 1,
   so this 'trick' gives them slightly more than the will intended.

11 Balancing

7T/16, 7C/12

This work introduces, in an informal way, the idea of doing the same thing to both sides of an equation.

	p 87	**A** Balance puzzles	Introduction to the idea of 'doing the same thing' to both sides of a balance
T	p 88	**B** Writing	Writing and solving a balance puzzle using a letter, for example $n + n + n = n + 10$
T	p 90	**C** Using shorthand	Using $3n$ notation instead of $n + n + n$

> **Essential**
> OHP
> Transparency of sheet 79
>
> **Practice booklet** pages 28 and 29

A Balance puzzles (p 87)

This section introduces the idea of scales balancing. The emphasis is on what can be done to both sides and still leave the scales in balance.

> An OHP, a transparency of sheet 79, cut up so that each picture is separate.

◊ Use the OHP pictures of weights and hedgehogs to discuss with the class what you can *add* to both sides of the scales and still keep them in balance. Point out that all the weights are (notionally) 1 kilogram each.

'This was excellent with an OHP.'

You could also ask the pupils what you can do to the scales and definitely make them unbalanced.

'I didn't have an OHP, so I did my own silly drawings on the whiteboard.'

Now discuss what you can take from both sides and still keep the scales in balance, finally finding out what one hedgehog weighs.

You can use the hedgehogs and rabbits to make up further problems for discussion. Include problems where there are weights on both sides.

A1–A3 Pupils are not expected to show any working when solving these puzzles. Of course, if they wish to write down any intermediate steps they should not be discouraged from doing so.

11 Balancing • 83

B Writing (p 88)

The emphasis here is on writing out the solution to the puzzle step by step. Pupils use the notation $n + n + n$ when solving problems. Section C introduces the shorthand $3n$ for $n + n + n$. Some pupils may suggest using this shorthand here, but do not force pupils into using the algebraic shorthand before they are ready for it.

> Optional: OHP and the cut-up transparencies used for section A.

◊ When introducing this section, point out that we are simply recording what is done to the balance in the most straightforward way possible. You could use the cut-up pictures and the OHP when doing this.

Emphasise that the unknown here (n in the example) stands for the *number of kilograms* each animal weighs. Avoid using letters like h (which pupils may think stands for the word 'hedgehog') or w (which pupils may think of as standing for 'a weight').

In your introduction, you may wish to include examples where weights and objects are on both sides of the balance. This is taken up in B4.

B3 Pupils choose their own letters to stand for the weight of each animal. Emphasise that the letter stands for the weight of the animal, not the name of the animal or the animal itself.

C Using shorthand (p 90)

This section introduces the notation $3n$ as a shorthand for $n + n + n$ when solving problems. Discuss the fact that, say, $8 + 8 + 8 = 3 \times 8$ and, in the same way, $n + n + n = 3 \times n$, which is written as $3n$.

A Balance puzzles (p 87)

A1 3

A2 2

A3 (a) 3   (b) 4   (c) 5

B Writing (p 88)

In this and the following section, pupils should write down their working and check.

B1 $n = 4$

B2 $x = 3$

B3 (a) 5   (b) 4   (c) 20

B4 $n = 2$ and check

B5 (a) 3   (b) 3

B6 The pupil's picture for
$s + s + s + 6 = s + 12$, with $s = 3$

B7 (a) $y = 12$ (b) $p = 3$ (c) $z = 2$
(d) $g = 2\frac{1}{2}$ (e) $n = 6$ (f) $b = 20$

C Using shorthand (p 90)

C1 (a) 4   (b) 8

C2 (a) $p = 5$ (b) $n = 5$ (c) $m = 3$
(d) $d = 6$ (e) $p = 7$ (f) $s = 8$

What progress have you made? (p 91)

1 4, with the pupil's working

2 (a) $w = 11$ (b) $d = 6$

3 (a) $m = 4$ (b) $y = 4$

Practice booklet

Sections A and B (p 28)

1 2

2 6

3 The pupil's working leading to
(a) 4   (b) 50   (c) 12   (d) 37

4 The pupil's working leading to
(a) 0.5 or $\frac{1}{2}$ (b) 12.5 or $12\frac{1}{2}$

5 The pupil's working leading to
(a) $s = 10$ (b) $w = 3$
(c) $t = 9$ (d) $p = 49$
(e) $t = 3$ (f) $u = 3$
(g) $v = 4$ (h) $w = 1\frac{1}{2}$
(i) $x = 33$

Section C (p 29)

1 The pupil's working leading to
(a) $t = 7$ (b) $w = 6$
(c) $m = 12$ (d) $t = 16$
(e) $p = 11$ (f) $x = 10$

⑫ Health club   7C/11

This practical introduction to some of the ideas of data handling is based on a set of data cards. The cards can be sorted, ordered, arranged to make bar charts, and so on.

A lot of ideas are introduced in this unit. They are all taken up again in more detail later in the course.

p 92	**A** On record	Using information on data cards
p 92	**B** On display	Median (odd number of values) Dot plots, grouped bar charts
p 94	**C** Males and females	Median (even number of values) Comparing two sets of data
p 94	**D** Two-way tables	

> **Essential**
>
> Sets of cards (one per pair of pupils) from sheets 80 and 81
> Sheets 82–87 (unless pupils draw their own axes for graphs)

🅐 On record (p 92)

This is to familiarise pupils with the data cards.

> A set of cards per pair of pupils (from sheets 80 and 81)

◊ You could start by asking pupils why a health club would keep the kind of information on the cards (which would of course have to be updated periodically). Explain that rest pulse (in beats per minute) is measured when the person is sitting still. It is a rough measure of fitness (the lower the better).

◊ Ask some questions to help familiarise pupils with the information on the cards, such as 'Who weighs 60 kg?' or 'Who is the tallest female?'

86 • 12 Health club

B **On display** (p 92)

This introduces grouped frequency bar charts, median and dot plots.

> The cards, sheets 82 and 83 (unless pupils draw their own axes for graphs)

◊ Ask pupils, in pairs, to sort the cards by age group. The cards can be placed to form a 'bar chart'. (You can do this on the board with Blu-tack.)

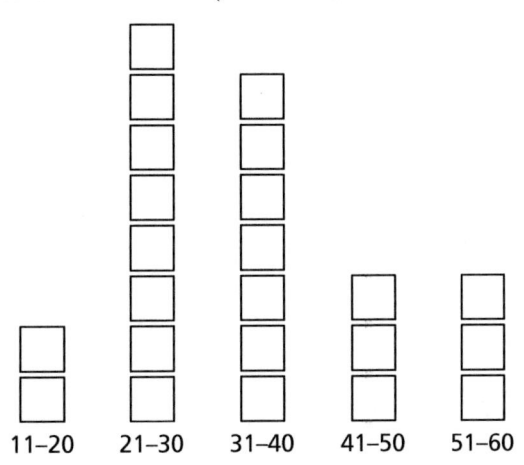

'I used a demo set of plastic coated cards. Pupils enjoyed putting them into grouped bar charts or on dot plots. Each child came and put a card on the board.'

11–20 21–30 31–40 41–50 51–60

Pupils then draw the bar chart (on sheet 82, if used).

Discuss the distribution of age groups and the possible reasons for this.

Median

◊ Ask pupils to look at the heights on the cards, to put the cards in order of height and to find the middle height. (Some pupils may think of this as halfway between the heights of the shortest and tallest people. You can say that this is one interpretation of 'middle height' but not what you meant.)

Emphasise that the median is a height (168 cm), not a person. (It helps to use 'median' always as an adjective before 'height', 'weight', etc.) Emphasise also that there are equal numbers of people shorter than and taller than the median height.

Dot plot

◊ The cards themselves can be used to illustrate a dot plot of the heights. Place them on a large number line using Blu-tack:

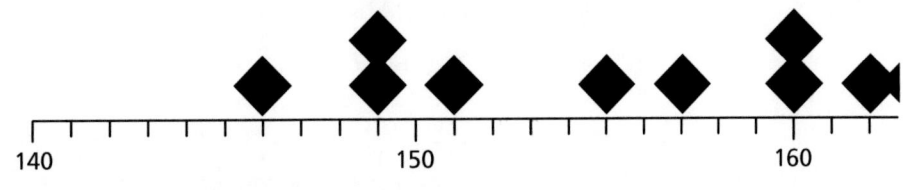

140   150   160

Pupils draw the dot plot on sheet 82.

12 Health club • 87

[Dot plot showing Height (cm) from 140 to 190]

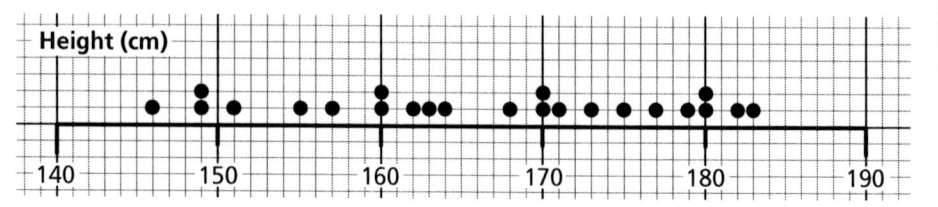

They could now discuss what the dot plot shows about the heights: they are fairly evenly spread but with more at the taller end.

Ask what sort of information can quickly be found from looking at the dot plot rather than the cards (for example, that the tallest person is 183 cm).

Grouped bar chart

◊ Compare the dot plot for heights with the bar chart for ages. The dot plot gives more detail but the bar chart shows the overall shape of the distribution better.

Discuss how to draw a bar chart for heights by grouping them.

Pupils draw a grouped bar chart on sheet 82. They can work directly from the cards or use the dot plot.

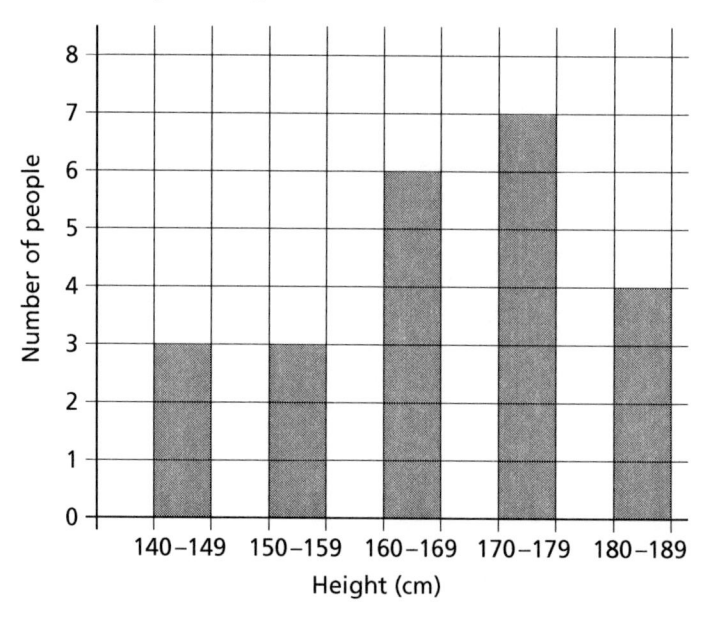

[Grouped bar chart: Number of people vs Height (cm). 140–149: 3, 150–159: 3, 160–169: 6, 170–179: 7, 180–189: 4]

The bar chart shows clearly that most people at the club are between 160 and 179 cm tall.

Some pupils could look at the effect of choosing different class intervals.

*B5 This raises the issue of the median of an even number of values.

88 • 12 *Health club*

C Males and females (p 94)

This is about comparing two sets of data.

> The cards, sheets 84–87

◊ Discuss the idea of comparing different groups of people. For example, the ages of males and females could be compared by drawing a bar chart for each (sheet 84).

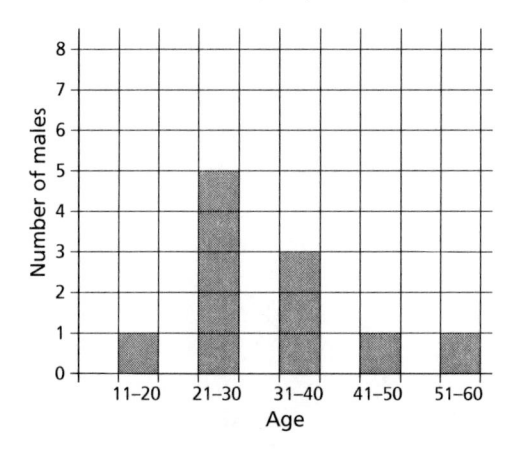

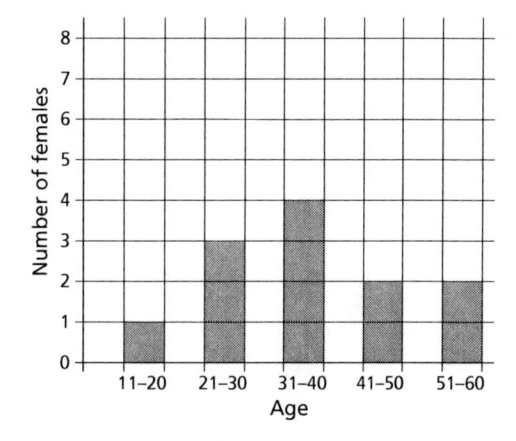

Comparing heights

◊ Discuss how the heights of the males and females could be compared. The median heights could be found. There are 12 females, so you will have to explain how to find the median in this case.

Finding the value halfway between two values is not as easy as it appears and the method of adding the two values and dividing by 2 can be baffling. It is more natural to halve the difference and add it to the smaller value. (A picture can help here.)

◊ Pupils can draw dot plots and grouped bar charts for the males and females on sheet 85.

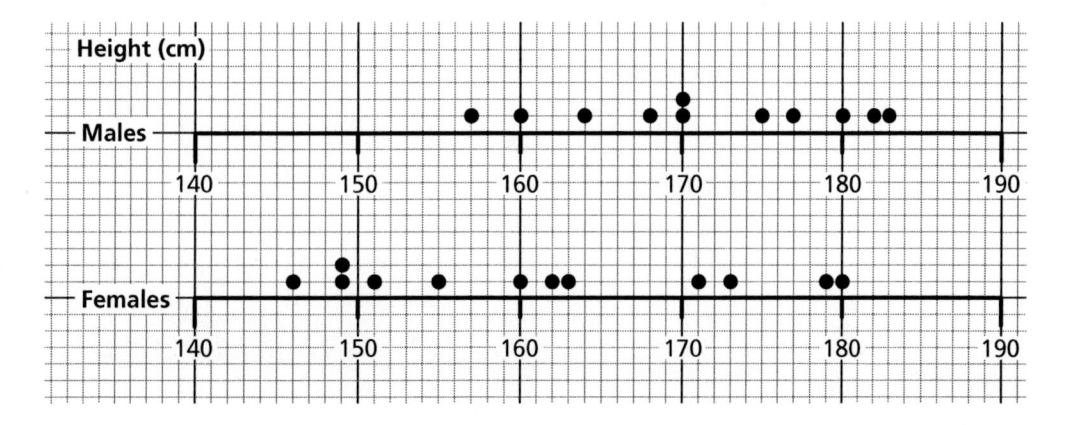

12 Health club • 89

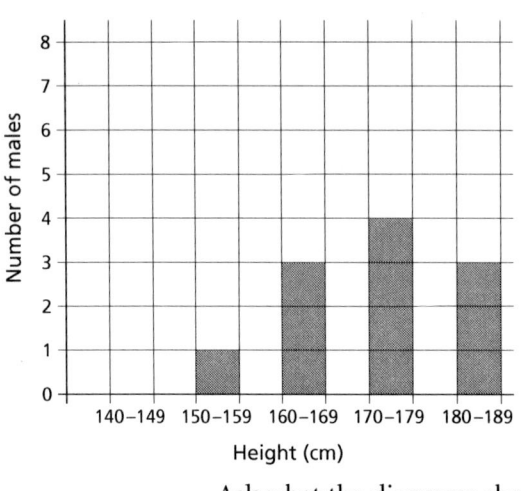

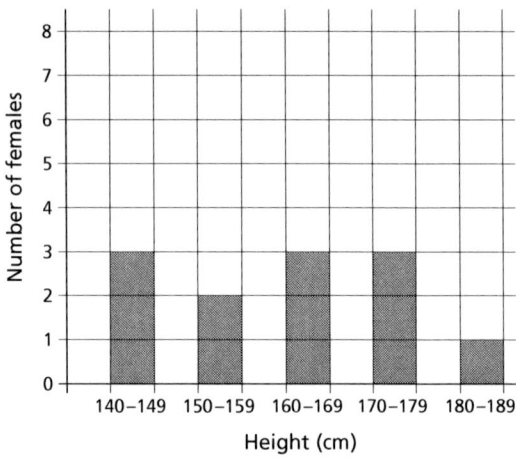

Ask what the diagrams show about the heights of males and females.

In one school, after working through the unit, pupils had a further lesson using a spreadsheet. The teacher had entered the data from the cards. Pupils were shown how to sort the data, for example in ascending order of rest pulse. They also used the median command.

They were shown how to filter the data to select out all the females and find median values for this subgroup. They then tackled questions C1 and C2 using the spreadsheet.

D **Two-way tables** (p 94)

> The cards

◊ The cards can be sorted into a large two-way diagram for example by age and gender:

	Age 11–40	Age 41–70
Male	▪ …	▪ ▪ …
Female	▪ ▪ ▪ …	▪ …

From this a two-way table of numbers can be made.

Further work

The cards can also be used to show a scatter diagram. For example, you could draw axes on the board for height and weight and Blu-tack each card at its position in the scatter diagram.

A On record (p 92)

A1 (a) 39 kg
(b) J Abram
(c) 8
(d) A Saunders, W Stanwell
(e) 13
(f) H Tear
(g) 5
(h) 2
(i) M Lakhani, G Peters

B On display (p 92)

B1 (a) 64 kg

(b)

(c)

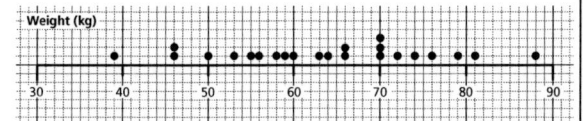

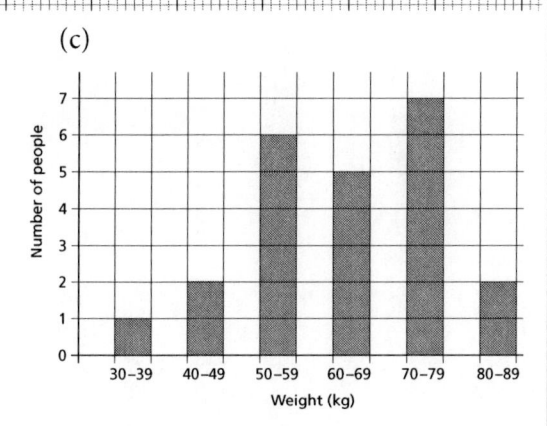

(d) The pupil's observations: for example, most people at the club weigh between 50 kg and 80 kg and not many people weigh over 80 kg.

B2 (a) 69 bpm

(b)

(c)

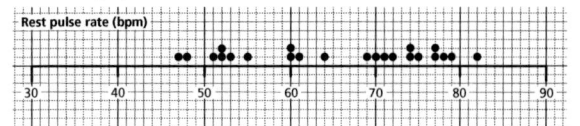

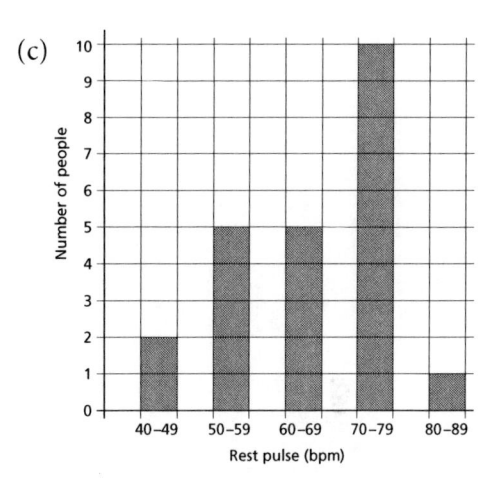

(d) The pupil's observations: for example, almost half of the people at the club have a rest pulse rate between 70 and 79 bpm.

***B3** (a) The actual ages are not recorded.
(b) 32 years and 39 years are possible median ages.

***B4** 22 years and 25 years are possible median ages.

***B5** (a) 167 cm
(b) 62.5 kg or $62\frac{1}{2}$ kg

C Males and females (p 94)

C1 (a) Males 70 kg, Females 57 kg
(b) The median weight for the males is 13 kg higher than the median weight for the females suggesting that the males at the health club tend to be heavier than the females.

12 Health club • 91

(c)

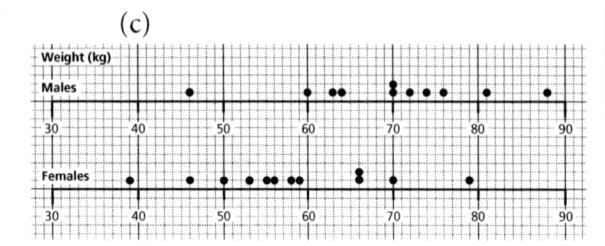

[Dot plot: Weight (kg) for Males and Females, scale 30–90]

(d)

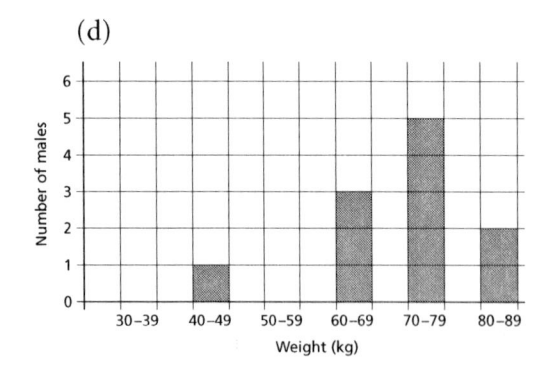

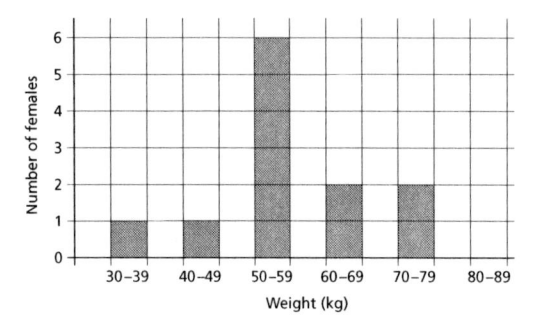

[Bar chart: Number of males vs Weight (kg), bars at 40–49 (1), 60–69 (3), 70–79 (5), 80–89 (2)]

[Bar chart: Number of females vs Weight (kg), bars at 30–39 (1), 40–49 (1), 50–59 (6), 60–69 (2), 70–79 (2)]

(e) The dot plots and bar charts support the claim that the men at the club tend to be heavier than the women. The dots for the males are further over to the right. On the bar chart for the males, the bars tend to be higher over to the right. The charts show that all the males except one weigh 60 kg or over but 8 out of the 12 females weigh below 60 kg.

C2 (a) The median rest pulse rate for the males is 69 bpm.
The median rest pulse rate for the females is 65 bpm.

(b) The median rest pulse rate for the males is only 4 bpm higher than the median rest pulse rate for the females suggesting that there is little difference between the rest pulse rates of the males and females at the health club.

(c)

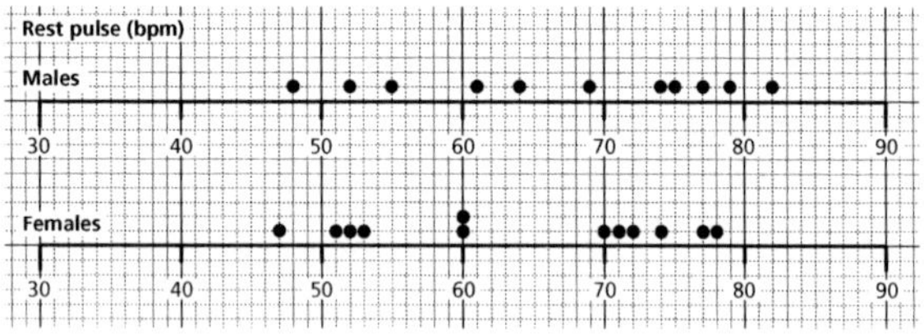

[Dot plot: Rest pulse (bpm) for Males and Females, scale 30–90]

(d)

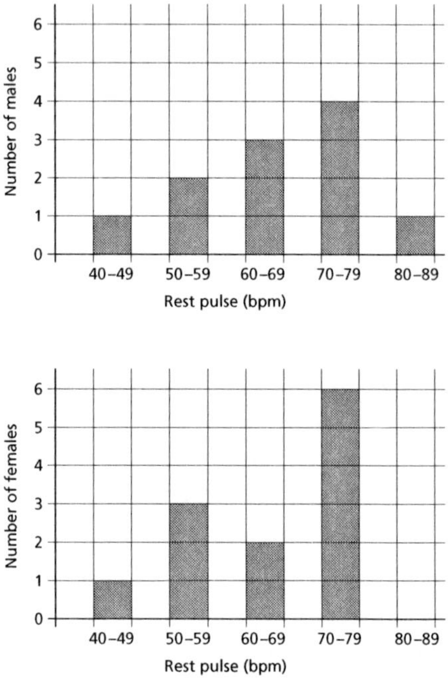

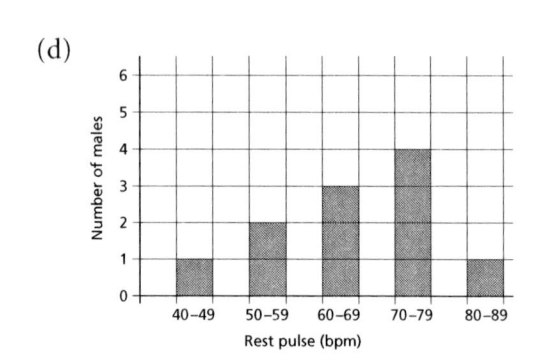

[Bar chart: Number of males vs Rest pulse (bpm), bars at 40–49 (1), 50–59 (2), 60–69 (3), 70–79 (4), 80–89 (1)]

[Bar chart: Number of females vs Rest pulse (bpm), bars at 40–49 (1), 50–59 (3), 60–69 (2), 70–79 (6)]

(e) The medians do not show a marked difference between the rest pulses for the males and females at the club. The dot plots show a fairly even spread for the males but a less even spread for the females where the dots tend to be clustered round two points. The bar chart for the females shows a 'dip' in the middle not shared by the graph for the males.

D Two-way tables (p 94)

D1

	Less than 150 cm	150 cm or more
Less than 50 kg	1	2
50 kg or more	2	18

D2

	Less than 60 bpm	60 bpm or more
11–40	3	14
41–70	4	2

13 Understanding decimals

7T/25, 7T/33

p 95   **A** Tenths of a centimetre

p 97   **B** Tenths

p 100 **C** Hundredths of a metre

p 102 **D** Up to two decimal places

p 105 **E** Up to three decimal places

p 106 **F** Multiplying and dividing by 10, 100, 1000, …

p 108 **G** Metric units

Essential	**Optional**
Sheets 40, 41, 42	Measuring tapes
Metre rules (for section C)	
Practice booklet pages 30 to 34	

A Tenths of a centimetre (p 95)

In this section, work on the first place of decimals is related to measurement in centimetres.

◊ Pupils commonly view 6.4 cm as 6 cm and 4 mm, just reading the number after the decimal point as mm (so 6.04 cm would be interpreted as 6 cm and 4 mm too!). Emphasise that we can write 6 cm and 4 mm as 6.4 cm because each millimetre is one tenth of a centimetre.

◊ You could ask pupils to decide if they think 2 cm and 2.0 cm are the same or different and to justify their decision.

B Tenths (p 97)

Sheet 40 (for B6), sheet 41 (for 'Decimal dominoes' and 'Getting in order')

◊ In your initial discussion pupils should
  • point to a range of decimals on the number line such as 2.5, 0.9 and 3.0
  • order a set of decimals such as 3, 2.9, 0.4, 4.0, using the number line to help if necessary
  • do mental calculations such as 1.5 + 0.1, 2 + 0.6, 3 − 0.5 and so on, using the number line to help if necessary

B4 This begins to address the misconception that, when reading scales, you count the number of divisions after the whole number to give the number after the decimal point.

B6 This question includes some experience of finding the position of decimals on a number line when divisions are not marked.

Target (p 98)

◊ You may prefer to postpone this game until later in the unit when pupils could include 2 decimal places.

Decimal dominoes (p 98)

These give practice in recognising that, for example, 1.9 differs from 2 by 0.1.

Getting in order (p 99)

◊ The cards include whole numbers as well as decimals, to give practice in an aspect which is often found more difficult.

◊ Another activity is to give each of, say, five pupils a sheet with a number. They have to get themselves in order as fast as they can.

C Hundredths of a metre (p 100)

Practical work may be needed for pupils who have previously recorded measurements in centimetres.

> Metre rules
> Optional: Measuring tapes

'In Technology pupils work in mm not cm. It was useful to explore this in class.'

◊ It is a good idea to rule columns on the board and record, say, pupils' heights in centimetres, in metres and centimetres, in metres and in millimetres. This will help them appreciate the relationships between the metric units.

◊ Show that the first figure after the decimal point shows tenths and the second figure the extra hundredths (a metre rule marked only in tenths of a metre could be helpful here).

◊ For practice in measuring and recording in metres, pupils can measure objects around the room and check each other's results.

13 Understanding decimals • 95

D Up to two decimal places (p 102)

> Sheet 42 (for D3)

◊ In your discussion, address common difficulties such as understanding the difference between, say, 2.8 and 2.08 and seeing that, say, 1.4 is larger than 1.32. Pupils also often find it difficult to see that, say, 0.32 is equivalent to 32 hundredths (as well as 3 tenths + 2 hundredths).

D3 This question includes some experience of finding the position of decimals on a number line when divisions are not marked.

E Up to three decimal places (p 105)

◊ Pupils can find it difficult to extend decimal notation beyond the second decimal place. A discussion of metres, centimetres and millimetres here may help.

F Multiplying and dividing by 10, 100, 1000, … (p 106)

◊ If pupils think 'to multiply by 10, add a zero', they will find this breaks down where decimals are involved. It is better to think in terms of place value: the figures move to the next higher place value when a number is multiplied by 10.

Similar comments apply to division.

G Metric units (p 108)

◊ Use the table to discuss how to convert metric quantities for weight and length, bringing out when to multiply and when to divide.

A Tenths of a centimetre (p 95)

A1 (a) 4.9 cm (b) 3.4 cm

A2 P 6.9 cm, Q 5.7 cm, R 8.9 cm

A3 (a) Key Q (b) Key R

A4 *a*: 5 cm or 5.0 cm   *b*: 5.4 cm
b is the longer line.

A5 Red line: 6.7 cm
Blue line: 6.1 cm
The red line is the longer line.

A6 Line *y* (3 cm) is 0.2 cm longer than line *x* (2.8 cm).

A7 (a) 0.9 cm (b) 0.4 cm (c) 0.6 cm

A8 (a) The pupil's estimate (b) 4.9 cm

A9 A: 3.3 cm (smallest)
B: 3.5 cm
C: 3.7 cm (largest)

B Tenths (p 97)

B1 (a) 3.1 (b) 2.4 (c) 2 (d) 3

B2 (a) 2.4 (b) 2.1 (c) 3.6 (d) 2.5
(e) 1.8 (f) 3.1 (g) 2.6 (h) 2

B3 (a) 0.9 (b) 0.6

B4 Debbie is right.

B5 (a) 8.3 (b) 9.5 (c) 10.2
(d) 0.5 (e) 2.5 (f) 4.5

B6 (a) The pupil's answers on sheet 40
(b) 0.9, 3.4, 7.1, 10.6
(c) The pupil's answers on sheet 40

B7 0.8 cm, 1.7 cm, 4.3 cm, 5 cm, 5.6 cm

B8 (a) 0.7, 1.5, 2.8, 3.4, 4
(b) 0.1, 0.4, 2, 2.9, 4.5
(c) 0.9, 1, 6, 7, 7.6, 8
(d) 0.7, 0.8, 1.2, 1.5, 2

B9 Kent (7 m), Jamal (6.8 m), Price (6.4 m), O'Brien (6.1 m), Stone (6 m)

B10 (a) 6 m
(b) Davis (6.3 m), Conrad (6.2 m), Perry (6 m)

C Hundredths of a metre (p 100)

C1 (a) 3.25 m (b) 4.68 m

C2 1.47 m

C3 (a) 1 metre and 98 centimetres
(b) 4 metres and 6 centimetres
(c) 9 metres and 80 centimetres
(d) 10 metres and 10 centimetres

C4 (a) 1.7 or 1.70 m (b) 1.07 m
(c) 2.45 m (d) 3.09 m

C5 (a) 2.38 m (b) 5.03 m
(c) 0.22 m (d) 6.3 or 6.30 m

C6 (a) 145 cm (b) 705 cm
(c) 360 cm (d) 5 cm

C7 1.18 m, 132 cm, 1.4 m, 1.66 m

C8 0.09 m, 0.53 m, 60 cm, 0.7 m

C9 Any height between 1.3 m and 1.4 m

C10 (a) Cobra (3.05 m)
(b) Boa constrictor (3.3 m)

C11 (a) 3.82 m (b) 3.88 m
(c) 3.90 m or 3.9 m (d) 3.94 m
(e) 1.25 m (f) 1.30 m or 1.3 m
(g) 1.38 m (h) 1.41 m
(i) 2.91 m (j) 2.96 m
(k) 3.01 m (l) 3.05 m
(m) 0.99 m (n) 1.01 m
(o) 1.05 m (p) 1.09 m
(q) 1.13 m

C12 4.12 m

C13 30 centimetres

C14 1.67 m

C15 18 centimetres

D Up to two decimal places (p 102)

D1 (a) 6.23 (b) 6.37 (c) 6.42
(d) 6.47 (e) 6.51 (f) 6.58
(g) 3.05 (h) 3.11 (i) 3.17
(j) 3.23

D2 (a) 0.86 (b) 0.91 (c) 0.99
(d) 1.08 (e) 1.15

D3 The pupil's answers on sheet 42

D4 (a) 8.25 (b) 4.95 (c) 2.45
(d) 6.05 (e) 3.85 (f) 10.05

D5 (a) 3.46 (b) 6.27
(c) 4.30 or 4.3

D6 You get 0.2 because the calculator leaves out the unnecessary zero.

D7 2.35 and 2.4

D8 5.17, 5.03 and 5.1

D9 (a) 4.76, 5.09, 5.32, 5.84, 6
(b) 2.8, 3.07, 3.19, 3.2, 3.5
(c) 0.15, 0.45, 0.6, 1.07, 1.1
(d) 0.46, 0.5, 0.92, 1.09, 1.3

D10 HOLIDAYS

D11 ENJOYABLE

A decimal puzzle: COMPLETED

D12 (a) 7 hundreds or 700
(b) 2 thousands or 2000
(c) 3 units or 3
(d) 1 ten or 10
(e) 8 hundredths or $\frac{8}{100}$ or 0.08

D13 (a) 2723.58 (b) 2714.58
(c) 2713.68 (d) 2713.59
(e) 2813.58

D14 5 stands for 500, 2 stands for 20, 3 stands for 0.3 or $\frac{3}{10}$, 4 stands for 0.04 or $\frac{4}{100}$. The zero indicates there are no units.

D15 (a) 44.6 (b) 55.6 (c) 45.5
(d) 45.61 (e) 7.54 (f) 7.39
(g) 5.34 (h) 7.24 (i) 68.74
(j) 28.84 (k) 20.74 (l) 403.86

D16 0.7

D17 0.03

D18 Two ways each from
(a) 5 units + 5 tenths + 8 hundredths
or 5 units + 58 hundredths
or 558 hundredths
(possibly also 55 tenths + 8 hundredths)
(b) 7 units + 5 tenths + 1 hundredth
or 7 units + 51 hundredths
or 751 hundredths
(possibly also 75 tenths + 1 hundredth)
(c) 1 unit + 8 hundredths
or 108 hundredths
(possibly also 10 tenths + 8 hundredths)
(d) 4 tenths + 2 hundredths
or 42 hundredths

D19 (a) 4.84 (b) 9.01 (c) 0.79 (d) 2.26

E Up to three decimal places (p 105)

E1 (a) 1.533 (b) 1.545 (c) 1.552
(d) 3.415 (e) 3.435 (f) 3.455

E2 (a) 3.005 (b) 0.538

E3 6.785

E4 (a) 1.287 (b) 0.306 (c) 5.984

E5 5.364, 5.37, 5.371, 5.379

E6 0.129, 0.13 (or 0.130), 0.131

E7 (a) 5.742 (b) 9.089 (c) 30.984

E8 (a) 1.183 (b) 2.165 (c) 0.005

E9 0.238

F Multiplying and dividing by 10, 100, 1000, ... (p 106)

F1 (a) 471 (b) 91.3 (c) 153.2
 (d) 90.7 (e) 20 954 (f) 9.85
 (g) 200.7 (h) 81 (i) 6430
 (j) 6780 (k) 840 (l) 59 100

F2 (a) 100 (b) 100 (c) 10 (d) 100
 (e) 10 (f) 100 (g) 100 (h) 1000
 (i) 10

F3 (a) 2.68 (b) 3.471 (c) 0.29
 (d) 2.098 (e) 1.49 (f) 8.3
 (g) 2.3 (h) 0.687 (i) 0.012
 (j) 0.015 (k) 8.591 (l) 0.0092

F4 (a) 10 (b) 100 (c) 100
 (d) 1000 (e) 100 (f) 100

F5 (a) 42 (b) 0.42 (c) 312.9
 (d) 634 (e) 21.819 (f) 4965
 (g) 0.132 (h) 510 (i) 0.294
 (j) 310 (k) 2 (l) 0.0086

F6 (a) 3.2 **320 32 0.32**
 (b) 0.6 **600 6 0.06**
 (c) 56 **0.056 0.0056 0.56**
 (d) **4.23** 423 **4.23 0.0423**

G Metric units (p 108)

G1 (a) B: 4.2 × 1000 (b) D: 4.2 ÷ 100

G2 (a) 1234 g (b) 2450 g (c) 1500 g
 (d) 800 g (e) 60 g

G3 (a) 23.9 km (b) 124.8 km (c) 0.56 km
 (d) 9.2 km (e) 0.09 km

G4 (a) 7.8 cm (b) 78 cm (c) 0.2 cm
 (d) 4510 cm (e) 9.8 cm

G5 (a) 0.6 m (b) 2350 m (c) 4.9 m
 (d) 90 m (e) 56.7 m

G6 67 g, 0.07 kg, 300 g, 0.5 kg, 892 g, 0.985 kg, 1.04 kg

G7 (a) 6000 ml (b) 3612 ml
 (c) 15 200 ml (d) 8070 ml
 (e) 20 ml

G8 (a) 4.765 litres (b) 6.7 litres
 (c) 0.9 litre (d) 0.042 litre
 (e) 0.007 litre

G9 250 ml

G10 7.5 kg

G11 40 days

G12 2500 days

What progress have you made? (p 109)

1 6.03

2 (a) About 24.5 cm (allow 24.3 to 24.7)
 (b) The pupil's table measurements

3 (a) 6.98 (b) 7.03 (c) 7.12

4 (a) 0.7, 1.9, 2.2, 4.3, 5
 (b) 0.09, 0.72, 0.8, 1.6, 1.76, 1.9

5 1.75

6 (a) 4.5 (b) 1.61 (c) 6.73
 (d) 28.92 (e) 342.1 (f) 0.186

7 (a) 5800 m (b) 7.61 m
 (c) 0.45 m (d) 0.067 m

Practice booklet

Sections A and B (p 30)

1 (a) The pupil's estimates
 (b) (i) 4.5 cm (ii) 6.4 cm
   (iii) 3.5 cm (iv) 5.7 cm

2 (a) 0.9 cm (b) 1.4 cm
 (c) 2.5 cm (d) 0.7 cm

3 (a) 6.9, 7, 7.1, 7.2
 (b) 6.2, 5.9, 6.1, 6

4 16.7, 16.8, 17.6, 18, 18.9, 19.6, 20, 23.1

5 (a) 0.8   (b) 3.5
6 (a) 14.3   (b) 15.8   (c) 16.2
   (d) 120.5   (e) 122.5
7 (a) 0.7, 1, 2.9, 3, 3.2, 4.6
   (b) 5, 5.9, 6.7, 7, 8.1, 10
   (c) 0.3, 1.1, 3.1, 3.5, 4, 5.9
   (d) 3.9, 4.8, 5, 10, 16.2, 17
8 1.9, 2, 2.1
9 B

Sections C, D and E (p 32)

1.
Mel	152 cm	**1 m 52 cm**	**1.52 m**
Ginger	**149 cm**	1 m 49 cm	**1.49 m**
Ali	**159 cm**	1 m 59 cm	1.59 m
Kay	**109 cm**	1 m 9 cm	**1.09 m**
Morag	**160 cm**	**1 m 60 cm**	1.6 m

2 (a) Toni
   (b) Jamie (1.44 m), Geri (1.5 m), Suki (1.58 m), Paul (1.6 m), Toni (1.62 m)
3 0.05 m, 0.5 m, 0.9 m, 1.01 m
4 B (1.57 m), C (153 cm)
5 (a) 3.94   (b) 3.99   (c) 4.02
   (d) 4.07   (e) 4.15   (f) 4.22
   (g) 4.26   (h) 4.31
6 4.25
7 (a) 2.07, 2.39, 2.5, 2.8
   (b) 1.01, 1.1, 1.43, 1.5
   (c) 4.09, 4.24, 4.25, 4.3, 5
   (d) 0.08, 0.65, 0.7, 1, 1.5
8 6.17, 6.3 and 6.37
9 0.01, 0.04, 0.2, 0.23, 0.4, 0.51, 0.6, 0.79
   giving TRIANGLE

10 (a) 63.7   (b) 72.7   (c) 62.9
   (d) 39.64   (e) 39.73   (f) 30.1
11 (a) 458.1 + **10** = 468.1
   (b) 34.91 + **0.01** = 34.92
   (c) 75.32 – **1** = 74.32
   (d) 54.8 – **0.8** = 54
   (e) 6.77 – **0.7** = 6.07
   (f) 6.77 – **0.07** = 6.7
12 (a) 0.68   (b) 3.71   (c) 4.40 or 4.4
   (d) 4.01   (e) 5.8   (f) 10.06
13 7.125
14 (a) 8.003   (b) 0.072
15 1.074

Sections F and G (p 34)

1 (a) 500   (b) 12   (c) 5.7
   (d) 35.4   (e) 3   (f) 2.32
   (g) 7684   (h) 2.43   (i) 0.345
   (j) 250   (k) 3450   (l) 0.81
   (m) 1249   (n) 6.426 (o) 23
   (p) 8.325 (q) 5020   (r) 0.037
   (s) 62.3   (t) 0.08   (u) 10
   (v) 150   (w) 20.05 (x) 0.000 34
2 (a) 100   (b) 10   (c) 6.2
   (d) 100   (e) 26   (f) 100
   (g) 1000   (h) 1.2   (i) 0.000 342
3 (a) 671 g   (b) 1230 g
   (c) 500 g   (d) 3 g
4 (a) 8.035 litres   (b) 0.5 litres
   (c) 0.839 litres   (d) 0.06 litres
5 (a) 4.3 cm   (b) 167 cm
   (c) 0.5 cm   (d) 320 cm
6 (a) 6300 m   (b) 60 m
   (c) 4.23 m   (d) 0.3 m
7 1.2 litres, 1.08 litres, 0.5 litres, 450 ml, 2.3 ml

14 Temperature

7T/31, 7C/18

T	p 110 **A** Colder and colder	Estimating and ordering temperatures Finding temperature differences
	p 112 **B** Temperature graphs	Reading a temperature graph
	p 113 **C** Graph or table?	Deciding which is easier to use
T	p 114 **D** Changes	Simple addition and subtraction of positive numbers

Essential
Sheets 90 to 92

Practice booklet pages 35 and 36

A Colder and colder (p 110)

Sheets 90 to 92

◊ You could start by asking pupils for examples of temperatures which they think they know. Can they give a reasonable estimate of the room temperature (about 20°C)? Have they seen different types of thermometer? Where?

◊ You could ask pupils to mark their suggestions for the temperatures A, B, C, … on a number line. Some pupils may have little idea of some of them.

'We had quite a heated (no pun intended) discussion on these and they were amazingly accurate in their final list.'

The approximate temperatures in order are:

B Human body temperature: 37°C
E Temperature of a heated swimming pool: 28°C
C Temperature of inside an ordinary fridge: 5°C
F Temperature of ice-cream when it's good to eat: 2°C
D Antarctic sea water temperature: ⁻1°C
A A winter's day temperature at the north pole: ⁻30°C

A discussion on the effect of salt on the freezing point of water may be needed for D.

More temperatures can be included in your discussion, for example:
- Oven temperature for baking a cake: 200°C
- Temperature of a hot bowl of soup: 55°C

- Temperature of a hot bath: 40°C
- Temperature at which butter melts: 35°C
- Temperature of a hot summer's day in Britain: 30°C
- Temperature inside a car in the morning after a frosty night: 1°C
- Temperature inside a home freezer: ⁻18°C

Temperature trumps (p 111)

This can be played in groups of 2, 3 or 4.
All the cards are dealt, face down. Players do not look at them.
The player whose turn it is reveals their top card and chooses either summer temperature or winter temperature.
The others turn over their top card.
If summer temperature was chosen then the highest summer temperature wins. If winter temperature was chosen then the lowest winter temperature wins.
The winner takes the turned-over cards and has the next turn.
The overall winner is the player with most cards at the end.

'Superb game! Worked well with all abilities.'

B **Temperature** (p 112)

C **Graph or table?** (p 113)

D **Changes** (p 114)

◊ Ask pupils how they see, for example, 2 − 5 on the number line. Help them to associate additions and subtractions with moves on the number line: 2 − 5 as 'start at 2, go down (or left) 5'.

◊ It is a good idea to distinguish between the negative sign ⁻ and the subtract sign −, at least to start with.

A Colder and colder (p 110)

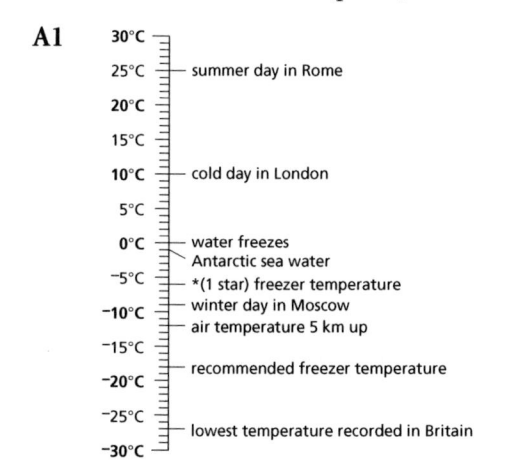

A1

```
30°C
25°C  — summer day in Rome
20°C
15°C
10°C  — cold day in London
 5°C
 0°C  — water freezes
      — Antarctic sea water
-5°C  — *(1 star) freezer temperature
      — winter day in Moscow
-10°C — air temperature 5 km up
-15°C
-20°C — recommended freezer temperature
-25°C
      — lowest temperature recorded in Britain
-30°C
```

A2 (a) −5°C (b) 3°C (c) −8°C

A3

	Highest	Lowest
(a)	2°C	−10°C
(b)	7°C	−9°C
(c)	−4°C	−8°C
(d)	−2°C	−10°C

A4 (a) 3°C (b) −1°C (c) −5°C
(d) −3°C

A5 (a) −15°C, −7°C, 0°C, 6°C, 14°C
(b) −18°C, −10°C, −3°C, 2°C, 8°C

A6 (a) 36°C (b) −18°C (c) −33°C
(d) 0°C

A7

	Highest	Lowest
(a)	1°C	−23°C
(b)	−5°C	−39°C

A8 29 degrees

A9 126 degrees

A10 (a) Scott (b) Nord
(c) 10 degrees (d) 9 months
(e) June, July, August
(f) November, December, January

B Temperature graphs (p 112)

B1 (a) 7°C (b) 8 p.m.
(c) −28°C (d) midnight
(e) 7°C (f) 16 degrees

B2 (a) −2°C (b) −13°C
(c) 11:30 p.m. and 1 a.m.
(d) $2\frac{1}{2}$ hours

C Graph or table? (p 113)

C1 (a), (b) Sturge
(c) Both table and graph are equally easy to use.

C2 Köge (Both table and graph are equally easy to use.)

C3 5 degrees
(The table is probably easier to use.)

C4 June, July, August
(The graph is probably easier to use.)

C5 November, December, January
(The table is easier to use.)

D Changes (p 114)

D1 (a) 1 (b) −4 (c) 0 (d) 2
(e) 3 (f) 5 (g) −2 (h) −1

D2 (a) 1 (b) −1 (c) −9 (d) −6
(e) −4 (f) −3 (g) −11 (h) −3

D3 (a) 2 (b) −10 (c) −1 (d) −8
(e) −4 (f) −4 (g) −3 (h) −13

D4 (a) 7 (b) −4 (c) 4 (d) −11

D5 (a) 3 (b) 4 (c) 5 (d) 2
(e) 1 (f) −2 (g) 1 (h) −1

D6 (a) 2 (b) −14 (c) 11 (d) −9
(e) 0

D7 (a) −7 (b) −16 (c) −27 (d) −10
(e) −80

D8 (a) 50 (b) 20 (c) 45 (d) ⁻40
(e) 115 (f) 121 (g) 60 (h) 48

D9 (a) 10, 8, 6, 4, **2, 0, ⁻2** (subtract 2)
(b) 7, 5, 3, 1, **⁻1, ⁻3, ⁻5** (subtract 2)
(c) 40, 30, 20, 10, **0, ⁻10, ⁻20** (subtract 10)
(d) ⁻2, ⁻4, ⁻6, ⁻8, **⁻10, ⁻12, ⁻14** (subtract 2)
(e) ⁻11, ⁻8, ⁻5, ⁻2, **1, 4, 7** (add 3)
(f) 24, 19, 14, 9, **4, ⁻1, ⁻6** (subtract 5)

D10 (a) Four out of these five calculations.
⁻4 + 6 = **2**   ⁻4 + 5 = **1**   ⁻4 + 3 = **⁻1**
⁻4 + 2 = **⁻2**   ⁻4 + 1 = **⁻3**

(b) The four calculations which are most likely to be made are
3 – **6** = **⁻3**   3 – **5** = **⁻2**   3 – **2** = **1**
3 – **1** = **2**

3 – ⁻2 = 5 and 3 – ⁻3 = 6 are unlikely to be found at this stage.

D11 (a) 5 (b) ⁻1 (c) 6 (d) ⁻5
(e) ⁻7 (f) ⁻10 (g) ⁻1 (h) ⁻10

D12 (a) 15 – 20 (b) ⁻5°C

D13 (a) ⁻20 + 38 = 18 (b) 40 – 42 = ⁻2
(c) ⁻60 + 35 = ⁻25 (d) 20 – 290 = ⁻270

D14 120°C

D15 (a) 5°C (b) 1000 m (c) 300 m
(d) 1300 m

What progress have you made? (p 117)

1 ⁻60°C ⁻18°C ⁻6°C 0°C 18°C

2 (a) ⁻10°C (b) 22°C

3 (a) Midnight and 8 a.m.
(b) 2 p.m.
(c) 8 p.m., about 9:20 a.m. and 6 p.m.

4 (a) 5 (b) ⁻3 (c) ⁻4 (d) ⁻14
(e) 6 (f) ⁻80 (g) ⁻47 (h) ⁻73

Practice booklet

Sections A and B (p 35)

1 ⁻5°C

2 (a) 4°C (b) ⁻2°C

3 (a) ⁻20°C, ⁻7°C, ⁻3°C, 1°C
(b) ⁻25°C, ⁻13°C, 4°C, 24°C

4 25 degrees

5 (a) ⁻9°C
(b) 4 a.m., 9 a.m., 4 p.m.
(c) ⁻13°C
(d) 4 hours
(e) 5 hours
(f) 7 degrees

Section D (p 36)

1 (a) 5 (b) ⁻6 (c) 0 (d) 16
(e) ⁻1 (f) ⁻8 (g) ⁻6 (h) ⁻23
(i) ⁻2 (j) ⁻14 (k) ⁻9 (l) 17

2 (a) ⁻2 + **10** = 8 (b) ⁻10 + **2** = ⁻8
(c) ⁻**15** + 5 = ⁻10 (d) ⁻1 + 1 = 0
(e) ⁻12 + **18** = 6 (f) ⁻**10** + 8 = ⁻2

3 (a) 10 – **4** = 6 (b) 3 – **7** = ⁻4
(c) 5 – **14** = ⁻9 (d) ⁻2 – **1** = ⁻3
(e) ⁻4 – **11** = ⁻15 (f) **3** – 4 = ⁻1
(g) **5** – 7 = ⁻2 (h) ⁻6 – 2 = ⁻8
(i) ⁻**17** – 3 = ⁻20

4 (a) ⁻27 + **33** = 6 (b) 18 – **30** = ⁻12
(c) 40 – **91** = ⁻51 (d) ⁻**20** + 32 = 12
(e) **3** – 13 = ⁻10 (f) **40** – 100 = ⁻60

5 (a) 3 (b) ⁻11 (c) ⁻5

6 (a) 180 – 210 = ⁻30; ⁻30°C
(b) 22 – 150 = ⁻128; ⁻128°C
(c) ⁻101 + 67 = ⁻34; ⁻34°C

15 Oral questions: money 1 (p 118)   7T/17

These pages are for teacher-led oral questioning, to develop mental skills with money in a 'Saturday market' context.

◊ Aim for sessions of oral questions that are regular and fairly short, with all pupils feeling a sense of achievement at the end. It is intended that you can use the pages more than once.

These sample questions are roughly in order of increasing difficulty.

1	What is the highest price marked on the stall?	£7.50
2	I have 50p. What can I afford to buy? (Invite alternatives.)	
3	What do a chocbar and a metal puzzle cost altogether?	90p
4	How much do four computer games cost?	£20.00
5	I buy a metal puzzle. How much change do I get from £1.00?	40p
6	How much do seven chews cost?	35p
7	What do a computer game and a CD cost?	£7.95
8	I buy a CD and give the stall owner £3.00. How much change do I get?	5p
9	What do a metal puzzle and a poster cost?	£1.10
10	How much do five chocbars cost?	£1.50
11	How much change do I get from a £5 note when I buy a set of Christmas tree lights?	75p
12	A woman has a £5 note. She buys a video tape. What else can she afford to buy? (Invite alternatives.)	
13	What do three cans of cola cost?	£1.35
14	How many posters can I buy for £2?	4
15	How much change do I get from a £5 note when I buy a pack of kitchen rolls?	£3.55
16	What do two video tapes cost?	£7.50
17	I buy a glue stick and a video tape. How much is that altogether?	£5.40
18	If computer games are reduced to half price how much will they be each?	£2.50
19	How much dearer is a video tape than a CD?	80p
20	If I buy two CDs, how much change do I get from a £10 note?	£4.10
21	Three friends buy a video tape and split the cost equally. How much does each friend pay?	£1.25
22	I have £5. How many glue sticks can I afford?	3
23	If I buy a T-shirt, a gent's tie, a pack of kitchen rolls and a glue stick, how much change do I get from £20?	£5.40
24	I have £10. If I buy two sets of Christmas tree lights, how many chews can I afford to buy?	30

16 Coordinates

7T/12

Essential	**Optional**
Sheets 67 and 69	OHP transparency of sheet 67
Practice booklet pages 37 and 38	

A Recording positions (p 120)

> Sheet 67 (an OHP is also useful)

◊ The two class games described below can be played at any time.

Coordinate bingo

Each pupil draws a grid labelled 0 to 5 on each axis. They mark seven grid points with little circles. This is their bingo card. You have a grid as well and call out points at random.

Four in a line

Draw a grid on the board labelled 0 to 6 on each axis. Choose two players who take turns to say the coordinates of a point. Label the points with the players' initials. The first to get four of their points in a line is the winner.

For a more demanding, but more interesting, game make it five in a line on a 0 to 9 grid.

B Digging deeper (p 122)

> Sheet 69 (includes $\frac{1}{2}$ and 0.5)

C Negative coordinates (p 123)

◊ The raised negative sign helps to distinguish between the number ⁻3 (negative 3 or minus 3) and the operation – 3 (subtract 3).

◊ You could play 'Coordinate bingo' again, with a grid labelled from ⁻3 to 2.

A Recording positions (p 120)

A1 (a) Boy's boot (b) Silver brooch (c) Wooden comb

A2 (a) (3, 7) (b) (7, 10) (c) (6, 3)

A3 A wall   **A4** (0, 0)

A5 A (3, 3), B (8, 8)

A6 Sheet 67

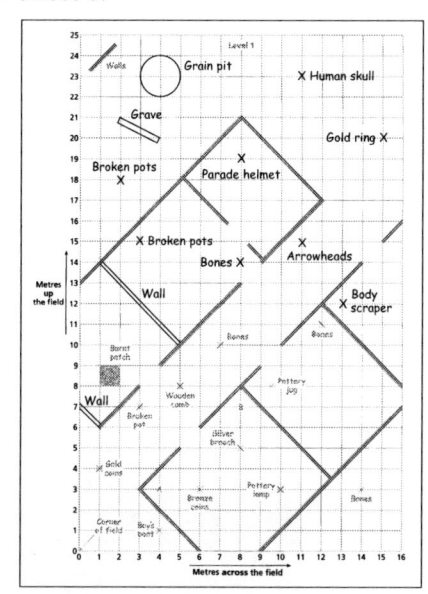

B Digging deeper (p 122)

B1 Sheet 69

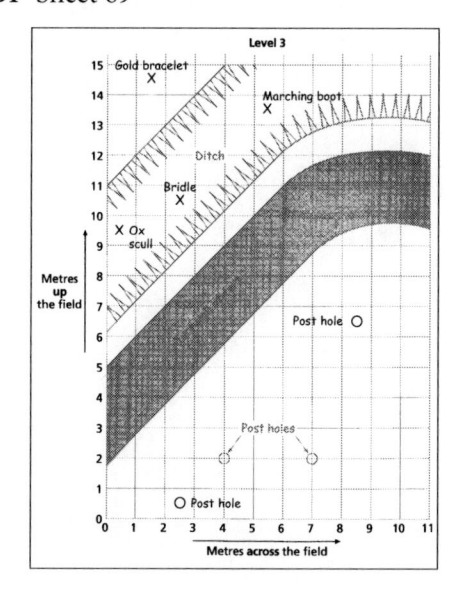

B2 Sheet 69

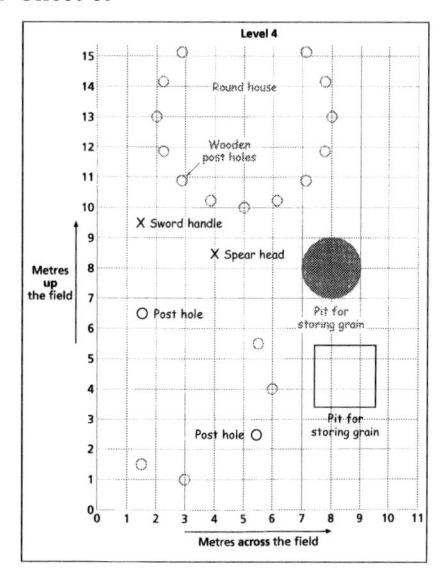

C Negative coordinates (p 123)

C1 (a) (⁻2, 2) (b) (4, ⁻2) (c) (⁻2, ⁻3)
      (d) (2, ⁻4) (e) (⁻4, 0) (f) (⁻6, 2)
      (g) (0, ⁻3)

C2

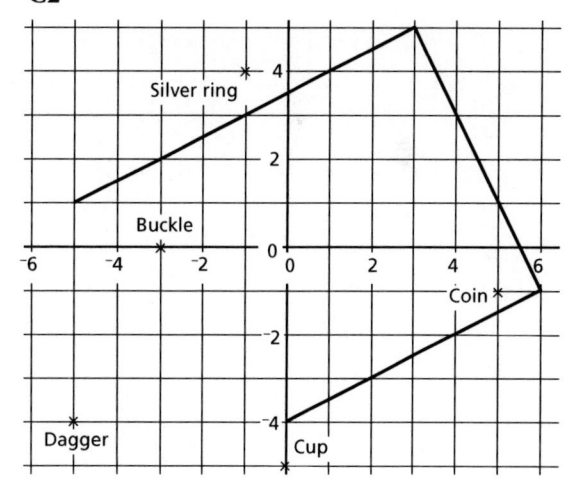

C3 (a) (⁻1, 4), (⁻1, ⁻2), (5, ⁻2)
      (b) (⁻1, 1), (2, ⁻2)

16 Coordinates • 107

What progress have you made? (p 124)

1 A (0, 3), B (4, 1), C (6, 0)

2

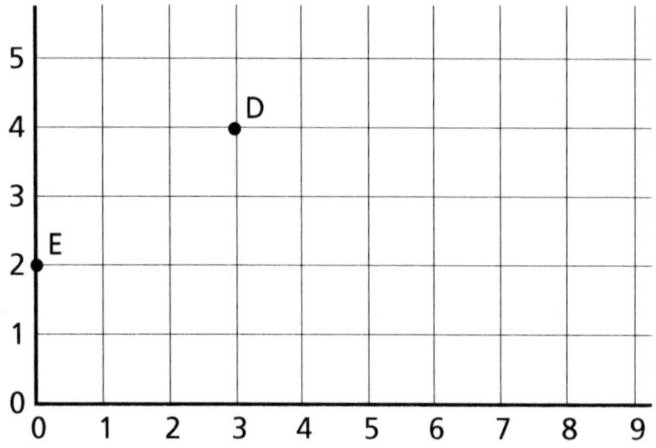

3 The points $(^-7, 2)$, $(^-3, ^-6)$ and $(1, ^-2)$ plotted on a grid

Practice booklet

Sections A and B (p 37)

1 (a) (0, 2), (3, 4), (6, 2), (3, 0)
(b) (3, 2)

2 (a) (4, 5), (4, 8), (7, 8), (7, 5)
(b) (5.5, 6.5)

3 (a) (7.5, 0.5), (7.5, 4.5), (9.5, 4.5), (9.5, 0.5)
(b) (8.5, 2.5)

4 (a) Inside A (b) Inside C
(c) Inside B (d) Not inside
(e) Inside A (f) Inside A

Section C (p 38)

1 (a) $(^-5, 3)$, $(1, 3)$, $(1, ^-3)$, $(^-5, ^-3)$
(b) $(^-2, 0)$

2 (a) $(8, ^-2)$ (b) $(5, ^-2)$

3

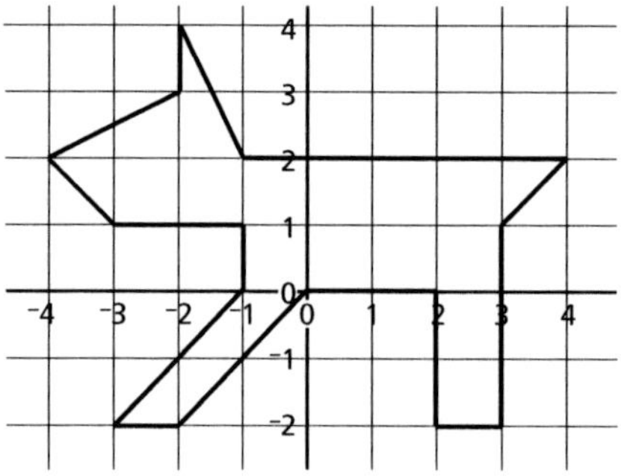

17 Number patterns

7T/26, 7C/6

p 125 **A** Exploring a number grid	
p 126 **B** Rectangles	Introducing prime numbers
p 127 **C** Squares, cubes and triangles	Square, cube and triangle numbers
p 129 **D** Sequences	

Optional
Sheet 96
Multilink or other cubes

Practice booklet pages 39 and 40

A Exploring a number grid (p 125)

These investigations are all based on the six-column grid and are graded in difficulty. Some teachers have preferred to do some or all of them later in the unit.

Optional: Sheet 96

Investigation 1 (Add a number in column A to one in column B)

◊ This is a good one to start on with the whole class. It leads on to variations which pupils can investigate for themselves.

The result of A + B is always in column C (provided the grid is extended downwards). You can then ask pupils what they think could be meant by 'investigate further'.

Here are some fruitful suggestions:

What if we add two different columns?
What if we subtract?
What if we multiply?

Investigation 2 (Multiples)

◊ These occur in spatially regular patterns.

Investigation 3 (Predict the 30th number in column B, etc.)

◊ There are various ways to do this. One is to work out the 30th number in column F (30 × 6 = 180) and then work back to 176.

Investigation 4 (Predict which column 500 will be in, etc.)

◊ 500 ÷ 6 = 83 remainder 2, so 500 will be in column B.

17 Number patterns • 109

Investigation 5 (Prime numbers)

◊ Prime numbers (except for 2 and 3) are only in columns A and E. This is because the numbers in the other columns are either even or multiples of 3.

B **Rectangles** (p 126)

Prime numbers are introduced as numbers which cannot be made into a rectangular array, only a single line.

C **Squares, cubes and triangles** (p 127)

> Optional: Multilink or other types of cube

D **Sequences** (p 129)

B **Rectangles** (p 126)

B1 3 × 8 (or 8 × 3). Don't count a single line (1 × 24) as a rectangle.

B2 10 × 2

B3 (a) 2 × 6, 3 × 4
(b) 2 × 8, 4 × 4
(c) 2 × 9, 3 × 6
(d) 2 × 15, 3 × 10, 5 × 6

B4 You can only make a single line with 17. Other numbers include 2, 3, 5, 7, …

B5 11, 13, 17

B6 Because even numbers can be made into a rectangle 2 × something

B7 (a) 11 (b) 17 (c) 19 (d) 23
(e) 29 (f) 31 (g) 37 (h) 43

C **Squares, cubes and triangles** (p 127)

C1 6 × 6 = 36 and 7 × 7 = 49

C2 64, 81, 100

C3 121, 144

C4 3, 5, 7, 9, … odd numbers

C5 (a) 16 (b) 25 (c) 121 (d) 400

C6 Rob has done 10 × 2. He should have done 10 × 10.

C7 (a) 13 (b) 33 (c) 73
(d) 61 (e) 155

C8 $6 = 2^2 + 1^2 + 1^2$
$7 = 2^2 + 1^2 + 1^2 + 1^2$
$8 = 2^2 + 2^2$
$9 = 3^2$
$10 = 3^2 + 1^2$
$11 = 3^2 + 1^2 + 1^2$
$12 = 2^2 + 2^2 + 2^2$
$13 = 3^2 + 2^2$
$14 = 3^2 + 2^2 + 1^2$
$15 = 3^2 + 2^2 + 1^2 + 1^2$
$16 = 4^2$
$17 = 4^2 + 1^2$
$18 = 3^2 + 3^2$
$19 = 3^2 + 3^2 + 1^2$
$20 = 4^2 + 2^2$
$21 = 4^2 + 2^2 + 1^2$
$22 = 3^2 + 3^2 + 2^2$
$23 = 3^2 + 3^2 + 2^2 + 1^2$
$24 = 4^2 + 2^2 + 2^2$
$25 = 5^2$
$26 = 5^2 + 1^2$
$27 = 5^2 + 1^2 + 1^2$ or $3^2 + 3^2 + 3^2$
$28 = 5^2 + 1^2 + 1^2 + 1^2$ or
$4^2 + 2^2 + 2^2 + 2^2$ or
$3^2 + 3^2 + 3^2 + 1^2$
$29 = 5^2 + 2^2$
$30 = 5^2 + 2^2 + 1^2$

C9 (a) 27 (b) 3^3

C10 64

C11 1, 8, 27, 64, 125

C12 (a) 72 (b) 150 (c) 152 (d) 657

C13 (a) 21, 28, 36
(b) Increase the amount added by 1.

C14 (a) 1, 9, 25, 36 (b) 1, 8
(c) 2, 3, 7 (d) 1, 3, 36

***C15** The sequence of square numbers:
1, 4, 9, 16, 25, …
The pupil's explanation

D Sequences (p 129)

D1 (a) 12 (b) Even numbers, or add 2

D2 (a) Add 4 (b) Subtract 3

D3 (a) 7, **11**, 15, **19**, 23, 27, **31**   add 4
(b) 1, **4**, 7, **10**, **13**, 16, **19**   add 3
(c) **41**, **37**, 33, **29**, 25, **21**, 17   subtract 4
(d) 44, **39**, 34, **29**, **24**, 19, **14**   subtract 5

D4 (a) Add 1, then 2, then 3 …
next number 22
(b) Add 2, then 4, then 6, then 8, …
next number 45
(c) Subtract 3, then 6, then 9, then 12,
… next number 22
(d) Add 5, then 7, then 9, then 11, …
next number 62
(e) Subtract 1, then 3, then 5, then 7, …
next number ⁻5

D5 $1 + 2 = 3$, $2 + 3 = 5$, $3 + 5 = 8$,
$5 + 8 = 13$, …

Challenge!

(a) 1, 2, 4, 8, 16, 32, **64**, **128**
double each time
(b) 1, **4**, **9**, 16, **25**, 36, 49, **64**
square numbers
(c) 2, 3, 5, 7, 11, 13, 17, **19**, **23**
prime numbers

What progress have you made? (p 130)

1 Second row × 2 is in first row.
First row × 2 is in second row.
Third row × 2 is in third row.

2 (a) 13 can only be divided by 1 and
by 13. It can't be arranged in a
rectangle, only a line.
(b) 23, 29

3 (a) 25 is 5 × 5 (b) 16 (c) 36

4 27

5 64

6 (a) The pupil's explanation
(b) 1, 3, 6

7 (a) 5, 9, 13, 17, 21, **25**, **29**
(b) 43, 37, 31, 25, 19, **13**, **7**
(c) **10**, **17**, 24, 31, 38, 45, **52**

8 (a) Subtract 6
Next numbers 22, 16
(b) Add 2, 4, 6, 8 … (add even numbers)
Next numbers 43, 57
(c) Multiply by 2
Next numbers 96, 192

Practice booklet

Sections A, B and C (p 39)

1 2 × 16 (or 16 × 2)

2 2 × 18, 3 × 12, 4 × 9, 6 × 6

3 No, 9 is not prime, with the pupil's explanation

4 (a) 7 (b) 23 (c) 41

5 (a) 9 (b) 49 (c) 81 (d) 100

6 (a) 7 (b) 16 (c) 37 (d) 36

7 (a) 24, 9, 27 (b) 16, 1, 9, 25

8 The pupil's diagram for 15 such as

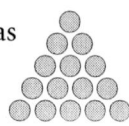

9 125

10 (a) 27 (b) 64 (c) 1000 (d) 8

11 $6^2 + 2^3$ (44), $10^2 - 6^2$ (64), 9^2 (81),
5^3 (125), $4^3 + 4^3$ (128), $3^3 + 5^3$ (152),
$10^3 - 9^2$ (919)

Section D (p 40)

1 (a) 25, 28 add 3
(b) 30, 26 subtract 4
(c) 80, 93 add 13
(d) 45, 34 subtract 11

2 (a) 28, 40 add 6
(b) 31, 34, 40 add 3
(c) 14, 32, 50 add 6
(d) 47, 44, 32 subtract 3

3 (a) Next two numbers 33, 45;
add 2, then 4, then 6, then 8, …
(b) Next two numbers 100, 136;
add 6, then 12, then 18, then 24, …
(c) Next two numbers 51, 66; add 3,
then 5, then 7, then 9, then 11, …
(d) Next two numbers 58, 44;
subtract 2, then 4, then 6, then 8,
then 10, …

18 Brackets

7T/19

The unit begins with a discussion of numerical expressions to help pupils see the need for brackets. It ends with 'Pam's game' which has been found successful in motivating pupils of all abilities. Algebraic expressions are not included.

You may have introduced brackets in earlier number work, for example 'Four digits' in 'First bites'. This unit should provide consolidation and extension.

You may wish to discuss the convention that multiplication and division take priority over addition and subtraction in expressions such as $3 + 4 \times 7$ and $16 - 10 \div 2$. However, this could be confusing for some at this stage and no use is made of it in the unit.

p 131 **A** Check it out	Using brackets to indicate which part of a calculation is to be done first
	Numerical expressions that use brackets
p 133 **B** Three in a row	Practising using one set of brackets
p 133 **C** Brackets galore!	Using more than one set of brackets
p 134 **D** Pam's game	Practising using brackets

Essential

Dice

Practice booklet page 41

A Check it out (p 131)

Teacher-led discussion introduces the idea that brackets indicate which part of a calculation is to be carried out first.

◊ Some of the expressions are correct as they stand, for example: $21 + 3$, $30 - 6$.

With some expressions, brackets can be used to indicate which part of the expression needs to be evaluated first, for example (Liz's second attempt): $(3 + 3) \times 4 = 24$ but $3 + (3 \times 4) \neq 24$.

A few of the expressions are equivalent to 24 with brackets in any position, for example $(2 \times 6) \times 2$ and $5 + 3 + 20 - 4$. There is no need to labour this point at this stage.

One expression is incorrect: $13 + 21$.

'We had already done "Four digits" and looked at brackets and calculators. This unit was a useful revision.'

◊ You can discuss how different calculators evaluate, say, 3 + 3 × 4, some working from left to right and others where multiplication and division take priority over addition and subtraction.

A5 Pupils can work in pairs or groups so solutions can be pooled and checked.

There are six different answers here but 24 calculations are possible (considering, for example, (2 + 5) × 3 and 3 × (2 + 5) to be different). Pupils could choose three numbers and two operations to give more or fewer than six answers.

A6 Digits should not be joined to make larger numbers (24, 46 etc.).

As an extension, pupils could investigate expressions that give the same value with brackets in any position. For example, (2 + 3) − 5 = 2 + (3 − 5) and (2 × 8) ÷ 4 = 2 × (8 ÷ 4).

They could try to find rules for when an expression is of this type.

B Three in a row (p 133)

This game consolidates the use of one set of brackets.

> Three dice for each pair/group (or one dice can be thrown three times), copies of the 'Three in a row' game board (pupils can copy this onto squared paper)

'Good fun, but weaker ones had to be supervised closely.'

◊ Pupils play the game in pairs or in larger groups (split into two teams).

◊ The squares of the game board can be made large enough to use counters.

C Brackets galore! (p 133)

Pupils work with multiple and nested sets of brackets.

◊ The '4s make 7' activity below can be used as a homework.

> **4s make 7**
>
> Make up as many expressions as you can which have a value of 7.
>
> You can use any of +, −, ×, ÷ , and brackets and 4 as often as you like.
>
> Here are some examples to start you off.
> (44 ÷ 4) − 4
> (4 + 4) − (4 ÷ 4)
> (4 + 4 + 4) ÷ 4) + 4
> (444 + 4) − (4 × 4 × 4) − 44 + 4

D Pam's game (p 134)

This game consolidates the use of brackets.

> One dice if game done as a class activity, otherwise enough dice for one for each group

◊ 'Pam's game' has been found successful in motivating pupils of all abilities to use brackets. It works well as a class activity or in small groups.

In one school, a teacher split her whole class into two opposing teams and set a time limit of 1 minute. Imposing a time limit may be necessary to keep the game moving – some pupils always want to get 100 exactly!

◊ Some pupils may adopt the convention that unless brackets show otherwise, an expression is evaluated from left to right. For example, 6 × (3 + 2) + (5 × 4) × 2 may be evaluated as 100 (6 × 5 + 20 × 2 worked from left to right). Some pupils will appreciate the need for extra brackets here to give ((6 × (3 + 2)) + (5 × 4)) × 2; others may not. It is likely you will want to emphasise this point more or less strongly to different groups of pupils.

◊ You could play with fewer than six numbers and/or a different target.

A Check it out (p 131)

A1 (a) 8 (b) 13 (c) 10 (d) 23
 (e) 20 (f) 2 (g) 20 (h) 0
 (i) 12 (j) 3 (k) 9 (l) 3

A2 (a) (6 − 1) × 3 = 15
 (b) 4 × (1 + 2) = 12
 (c) (2 + 1) × 5 = 15
 (d) (6 ÷ 3) + 9 = 11
 (e) 2 + (3 × 4) = 14
 (f) 5 × (2 − 1) = 5
 (g) (5 − 1) × 4 = 16
 (h) 2 + (2 × 2) = 6
 (i) 3 × (3 − 3) = 0
 (j) (4 + 4) ÷ 4 = 2
 (k) 12 ÷ (3 × 2) = 2
 (l) 10 − (6 − 2) = 6

A3 A and Y, B and Z, C and X

A4 (a) (1 + **4**) × 2 = 10
 (b) (**5** − 2) × 4 = 12
 (c) (3 × **3**) − 5 = 4
 (d) 2 × (10 − **8**) = 4
 (e) 4 × (**13** − 3) = 40
 (f) (6 + **9**) ÷ 3 = 5
 (g) 9 ÷ (**5** + 4) = 1
 (h) 10 ÷ (6 − **1**) = 2
 (i) 20 ÷ (**4** + 6) = 2
 (j) (4 × 3) − (5 − **2**) = 9

A5 (2 × 3) + 5 = 11
 (2 × 5) + 3 = 13
 (3 × 5) + 2 = 17
 2 × (3 + 5) = 16
 3 × (2 + 5) = 21
 5 × (2 + 3) = 25

There are only these six different numbers, but 24 calculations are possible.

18 Brackets • 115

A6 Examples are

6 − (4 − 2)   6 − (4 ÷ 2)
(6 − 4) × 2   2 × (6 − 4)
(6 − 4) + 2   2 + (6 − 4)
(6 + 2) − 4

C **Brackets galore!** (p 133)

C1 (2 + 3) × (10 − 3) = 35
((4 × 20) ÷ 2) × 5 = 200
(10 − (2 + 5)) × 3 = 9
8 + (2 × (4 + 6)) = 28

C2 (a) There are four different possible values:
(5 + 3) × (4 − 1) = 24
((5 + 3) × 4) − 1 = 31
(5 + (3 × 4)) − 1 = 16 or
5 + ((3 × 4) − 1) = 16
5 + (3 × (4 − 1)) = 14

(b) There are five different possible values:
(12 ÷ 2) + (4 × 2) = 14
((12 ÷ 2) + 4) × 2 = 20
(12 ÷ (2 + 4)) × 2 = 4
12 ÷ ((2 + 4) × 2) = 1
12 ÷ (2 + (4 × 2)) = 1.2

C3 There are six different possible values (one of which is negative):
((9 × 4) ÷ (2 × 3)) − 1 = 5 or
(9 × (4 ÷ (2 × 3))) − 1 = 5

((9 × 4) ÷ 2) × (3 − 1) = 36 or
(9 × (4 ÷ 2)) × (3 − 1) = 36 or
9 × ((4 ÷ 2) × (3 − 1)) = 36

(9 × 4) ÷ (2 × (3 − 1)) = 9 or
9 × (4 ÷ (2 × (3 − 1))) = 9

(((9 × 4) ÷ 2) × 3) − 1 = 53 or
((9 × (4 ÷ 2)) × 3) − 1 = 53 or
(9 × ((4 ÷ 2) × 3)) − 1 = 53 or
(9 × (4 ÷ 2) × 3) − 1 = 53

(9 × 4) ÷ ((2 × 3) − 1) = 7.2 or
9 × (4 ÷ ((2 × 3) − 1)) = 7.2

9 × ((4 ÷ (2 × 3)) − 1) = ⁻3

9 × (((4 ÷ 2) × 3) − 1) = 45

What progress have you made? (p 134)

1 (a) 15   (b) 10   (c) 10
(d) 20   (e) 2   (f) 14

2 (a) (6 + 1) × 2 = 14
(b) 3 × (10 − 8) = 6
(c) 10 − (6 ÷ 2) = 7

Practice booklet

Sections A, B and C (p 41)

1 (a) 11   (b) 16   (c) 8   (d) 3
(e) 4   (f) 30   (g) 6   (h) 4
(i) 12

2 (a) 4 × (3 + 2) = 20
(b) (4 × 3) + 2 = 14
(c) 10 − (3 × 2) = 4
(d) (10 − 3) × 2 = 14
(e) (4 + 5) × 2 = 18
(f) 4 + (5 × 2) = 14
(g) (12 ÷ 3) + 1 = 5
(h) 12 ÷ (3 + 1) = 3
(i) 9 + (4 ÷ 2) = 11

3 (a) $(2 \times \mathbf{3}) + 1 = 7$
 (b) $(3 + \mathbf{2}) \times 2 = 10$
 (c) $\mathbf{3} + (4 \times 2) = 11$
 (d) $9 - (\mathbf{3} + 2) = 4$
 (e) $(\mathbf{6} \div 2) + 5 = 8$
 (f) $12 \div (1 + \mathbf{5}) = 2$
 (g) $2 \times (8 - \mathbf{5}) = 6$
 (h) $10 - (\mathbf{7} - 3) = 6$

4 There are five different possible numbers:

$4 + (2 \times 3) = 10$ (the example)

$(4 + 2) \times 3 = 18$

$2 + (3 \times 4) = 14$

$(2 + 3) \times 4 = 20$

$3 + (4 \times 2) = 11$

These results can be found in different ways, for example $(4 + 3) \times 2 = 14$

5 (a) 4 (b) 10 (c) 0
 (d) 6 (e) 10

6 (a) $(12 + 6) \div (3 - 1) = 9$
 (b) $(12 + (6 \div 3)) - 1 = 13$
 (c) $((12 + 6) \div 3) - 1 = 5$
 (d) $12 + (6 \div (3 - 1)) = 15$

Review 2 (p 135)

> **Essential**
> Centimetre squared paper

1 $x + x + x + x + x + 3 = x + x + x + 15$
or $5x + 3 = 3x + 15$
with working leading to $x = 6$

2 (a) 6 (b) 20 (c) 6
(d) 12 (e) 15

3 (a)

[Graph showing rhombus with vertices A(6,2), B(4,4), C(2,2), D(4,0) on grid 0–7 × 0–7]

(b) 3.6 cm (c) (5, 1)

4 (a) −10°C, −6°C, 0°C, 2°C
(b) −15°C, −4°C, −3°C, 7°C

5 (a) +5 (b) 22, 27, 32

6 (a) $2 \times (5 + 3) = 16$
(b) $(20 - 10) \div 2 = 5$
(c) $20 - (10 \div 2) = 15$

7 (a) $\frac{2}{8}$ or $\frac{1}{4}$ (b) $\frac{4}{8}$ or $\frac{1}{2}$ (c) $\frac{3}{4}$

8 36 cm

9 (a) 2 (b) −7 (c) −4 (d) −5

10 (a) $u = 4$ (b) $v = 3$

11 (a) 1.06 m (b) Yes

12 (a) 4.96 (b) 5.04 (c) 5.15

13 17

14 (a) 3 (b) 8 (c) 6

15 23 mm, 1.08 m, 1.2 m, 1.46 m, 1.5 m, 198 cm, 2 m

16 (a) +4 (b) −5, −1, 3

17 (a) 3.85 (b) 19.71 (c) 2.882

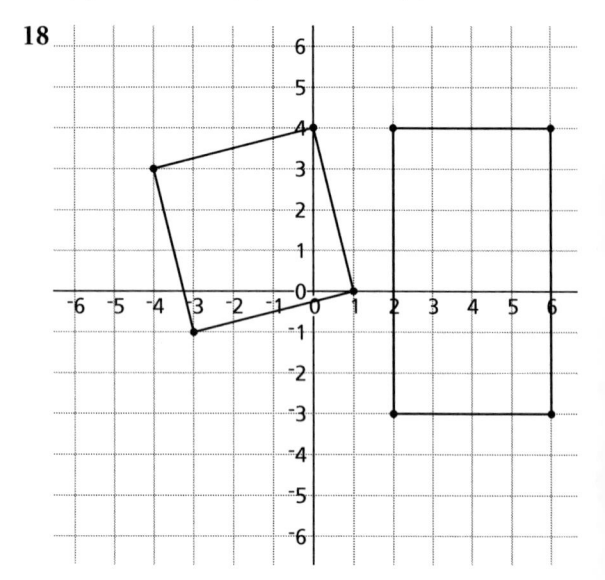

18

[Coordinate grid showing a square rotated and a rectangle]

(a) Fourth point (2, −3)
(b) Fourth point (0, 4)

19 (a) 12 (b) 4.5 (c) 13 500
(d) 0.014

20 (a) $\frac{4}{12}$ or $\frac{1}{3}$ (b) $\frac{5}{12}$ (c) 4

21 20 days

22 (a) 49 (b) 125 (c) 29
(d) Two from 1, 3 and 6

23 86 cm

Mixed questions 2 (Practice booklet p 42)

1 (a) $\frac{1}{2}$ (b) $\frac{1}{3}$ (c) $\frac{3}{8}$ (d) $\frac{5}{8}$

2 $n + n + n = n + 28$
with working leading to $n = 14$

3 (a) 47.7 (b) 48.9 (c) 50.5

4 −9°C

5 (a) 10 (b) 14 (c) 20

6 (a) 12   (b) 4   (c) 10   (d) 20
    (e) 16   (f) 40   (g) 36   (h) 27

7 (a) $r = 8$ (b) $m = 11$

8 David (98 cm), Col (1 m 9 cm), Eric (1.1 m), Vicky (1.25 m), Di (1.3 m)

9 (a) 12°C is **2** degrees higher than 10°C.
    (b) ⁻2°C is 8 degrees higher than ⁻10°C.
    (c) ⁻8°C is 4 degrees higher than **⁻12**°C.
    (d) ⁻4°C is 4 degrees **lower** than 0°C.

10 6.9, 7, 7.25, 7.28, 7.3, 7.4, 7.41

11 (a) Points P (⁻2, ⁻3), Q (⁻4, 1) and R (0, 3) plotted on a grid
    (b) (2, ⁻1)   (c) (⁻1, 0)

12 (a) 37   (b) 4.5   (c) 7.09
    (d) 1000   (e) 0.43   (f) 150

13 (a) 25   (b) 23

14 (a) 54.78   (b) 2.29   (c) 19.76

15 (a) 6   (b) 4   (c) 2

16 7.6 cm

17 76

19 Rounding decimals

A calculator is to be used where appropriate.

p 137 **A** Nearest whole number

p 139 **B** Rounding to one decimal place

p 140 **C** Rounding to two decimal places

p 142 **D** Rounding 'long' decimals

> **Optional**
> Metre rules
>
> **Practice booklet** pages 44 and 45

A Nearest whole number (p 137)

Pupils round numbers with one or two places of decimals to the nearest whole number.

Optional: Metre rules

◊ Discuss how the newspapers have rounded the figures in their articles to give the numbers in the headlines. Centimetre rulers, metre rules and number lines can all help pupils' understanding. Discussion of rounding amounts of money to the nearest pound can help in the discussion of how to round numbers with two decimal places.

B Rounding to one decimal place (p 139)

Pupils round numbers with two or three places of decimals to one decimal place.

Optional: Metre rules

◊ As before, metre rules and number lines can help pupils' understanding. Pupils usually find it harder to see that, for a number with three places of decimals, you only need to look at the second digit to decide whether to round up or down. You may wish to have two discussion sessions, one before questions B1 to B5 and one before B6 to B11.

◊ Towards the end of your discussion, include examples of the more difficult types such as 1.02 and 4.98.

B11 As an extension to this question, pupils could make up similar problems for others to solve.

C **Rounding to two decimal places** (p 140)

Pupils round numbers with three or four places of decimals to two decimal places.

Optional: Metre rules

◊ As before, you may wish to have two discussion sessions, one before questions C1 to C5 and one before C6 to C14.

D **Rounding 'long' decimals** (p 142)

◊ Pupils can work on the initial problems in pairs or groups and discuss their conclusions with the rest of the class. Pupils usually find it difficult to grasp the relative size of what each digit represents in the later decimal places. A diagram such as the one below can help.

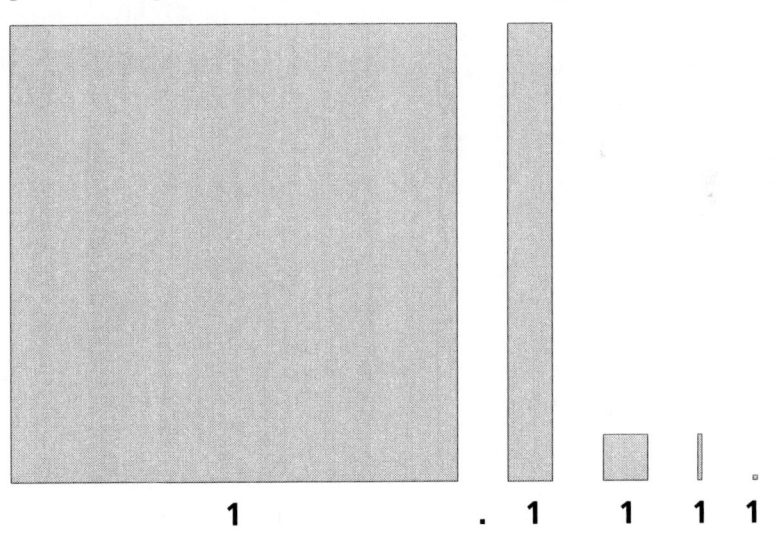

1   . 1   1   1   1

A **Nearest whole number** (p 137)

A1 4

A2 (a) 3   (b) 4   (c) 3

A3 (a) 7 cm   (b) 8 cm

A4 (a) 7   (b) 5   (c) 12   (d) 16
(e) 20   (f) 40   (g) 8   (h) 1
(i) 120   (j) 300

A5 (a) 4   (b) 18   (c) 20
(d) 32   (e) 150

19 Rounding decimals • 121

A6 (a) 3 (b) 4 (c) 6
(d) 9 (e) 10

A7 4

A8 (a) 3 (b) 4 (c) 3
(d) 4 (e) 3 (f) 4

A9 6 kg

A10

Zig Zag	£12
Arrow	£14
Blazer	£16
Astro	£21
Rapido	£25
Cruiser	£35
Sky Racer	£46

A11 (a) 1 m (b) 15 m (c) 28 m
(d) 30 m (e) 1 m

A12 (a) 16 (b) 8 (c) 7
(d) 291 (e) 10

A13 (a) 2 kg (b) 3 kg (c) 4 kg
(d) 14 kg (e) 10 kg

B Rounding to one decimal place
(p 139)

B1 2.8 kg

B2 (a) 2.9 (b) 2.8 (c) 2.9 (d) 2.8
(e) 2.8 (f) 2.9 (g) 2.8 (h) 2.9
(i) 2.8 (j) 2.9

B3 (a) 3.6 (b) 9.3 (c) 7.6 (d) 4.2
(e) 6.1 (f) 0.8 (g) 0.3 (h) 9.1
(i) 8.9 (j) 0.1

B4 (a) 42.7 m (b) 23.3 m (c) 19.4 m
(d) 2.9 m (e) 8.1 m

B5 (a) 5.0 (b) 2.0 (c) 5.0 (d) 6.4
(e) 1.8 (f) 16.8 (g) 7.5 (h) 7.5
(i) 4.1 (j) 1.0

B6 2.8

B7 (a) 2.8 (b) 2.9 (c) 2.9 (d) 2.9
(e) 2.8

B8 (a) 4.6 (b) 1.2 (c) 7.4 (d) 12.3
(e) 2.8 (f) 16.2 (g) 13.8 (h) 2.0
(i) 1.1 (j) 0.9

B9 (a) 1.6 cm (b) 0.7 cm (c) 0.7 cm
(d) 4.0 cm (e) 1.3 cm

B10 (a) 1.3 (b) 5.9 (c) 4.4
(d) 10.4 (e) 0.5

B11 (a) 1.7, 1.4, 1.2, 1.0 → ERBA → BEAR
(b) 1.3, 1.4, 1.0, 1.2, 1.5, 1.2 → TRABIB → RABBIT
(c) 1.4, 1.5, 1.3, 1.7, 1.6 → RITEG → TIGER
(d) 1.1, 0.9, 1.8, 1.4, 1.3, 1.7, 1.0 → MHSRTEA → HAMSTER

C Rounding to two decimal places
(p 140)

C1 (a) 2.86 (b) 2.87 (c) 2.87
(d) 2.86 (e) 2.87

C2 (a) 3.74 (b) 3.79

C3 (a) 3.71 (b) 3.79 (c) 3.77
(d) 3.80 (e) 3.70

C4 (a) 4.68 (b) 7.81 (c) 1.63
(d) 36.20 (e) 0.19 (f) 14.61
(g) 0.03 (h) 9.02 (i) 11.41
(j) 19.40

C5 (a) 5.94 m (b) 0.24 m (c) 3.46 m
(d) 2.50 m (e) 1.01 m

C6 (a) 2.87 (b) 2.86 (c) 2.86
(d) 2.86 (e) 2.87

C7 B (1.3563), D (1.3612), E (1.3609)

C8 (a) 3.48 (b) 1.56 (c) 12.92
(d) 8.94 (e) 1.06 (f) 0.98
(g) 405.72 (h) 0.05 (i) 11.40
(j) 0.80

C9 (a) £9.88 (b) £1.54 (c) £0.44
(d) £10.86 (e) £3.24 (f) £0.19
(g) £5.10 (h) £18.71 (i) £4.60
(j) £0.10

C10 19 g

C11 £0.58 or 58p

C12 1.29 kg

C13 21.9 cm

*C14 (a) 5.94 m (b) 6 cm
(c) 20 mm or 2 cm
(d) 7 cm (e) 0.60 m

D Rounding 'long' decimals (p 142)

D1 11.1 cm

D2 (a) 8.2 (b) 10.7 (c) 2.3 (d) 0.4
(e) 34.0 (f) 3.8 (g) 3.3 (h) 1.2

D3 (a) 4 g (b) 183 g

D4 16.7 cm

D5 £0.21 or 21p

D6 Family pack: £0.16 or 16p
Large pack: £0.12 or 12p
Giant pack: £0.11 or 11p
Jumbo pack: £0.10 or 10p

D7 Light: £0.08 or 8p
Low-fat: £0.20 or 20p

D8 (a) 0.77 (b) 6.69 (c) 2.07 (d) 28.85

What progress have you made? (p 143)

1 £2

2 (a) 8 (b) 12 (c) 40
(d) 4 (e) 1 (f) 106

3 (a) 3.4 (b) 6.6 (c) 1.4
(d) 5.0 (e) 1.3 (f) 16.8

4 6.5 cm

5 (a) 1.26 (b) 6.02 (c) 0.32

6 £0.86 or 86p

Practice booklet

Sections A, B and C (p 44)

1 5 cm

2 The pupil's table with these weights (in kg) in the second column: 70, 73, 75, 78, 81, 83, 84, 83

3 (a) 1 m (b) 7 m (c) 4 m
(d) 14 m (e) 2 m

4 (a) 8.5 (b) 0.9 (c) 7.8
(d) 10.1 (e) 4.3 (f) 11.3
(g) 15.4 (h) 0.4 (i) 21.5
(j) 20.1

5 (a) 7.27 (b) 3.61 (c) 42.74
(d) 49.20 (e) 12.49 (f) 0.20
(g) 36.10 (h) 48.20 (i) 4.35
(j) 19.06

6 £0.40 or 40p

7 0.93 m

8 3.4176, 3.4225

9 5.67

10 9.1

Section D (p 45)

1 £0.12 or 12p

2 Family pack 16p
Large pack 12p
Giant pack 11p

3 157 g

4 (a) 20.9 (b) 128.0 (c) 1.9
(d) 0.5 (e) 152.0 (f) 4.3
(g) 1.1 (h) 0.9

5 (a) 0.74 (b) 15.27 (c) 33.06
(d) 0.14

20 Gravestones

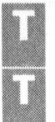

7T/18, 7C/16

T	p 144 **A** What gravestones tell us	Reading from a table
T	p 145 **B** Making a frequency table	Grouping, tallying
	p 146 **C** Comparing charts	Ways of showing frequencies
T	p 148 **D** Charts galore!	Good and bad charts
T	p 150 **E** Testing a hypothesis	
	p 151 **F** Modal group	

> **Essential**
> Sheet 93
> **Practice booklet** pages 46 and 47

A What gravestones tell us (p 144)

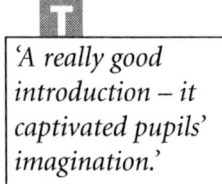

'A really good introduction – it captivated pupils' imagination.'

◊ In some circumstances, such as a recent bereavement, sensitive handling is needed.

The topic obviously lends itself to locally based practical work. The sample you get from a graveyard is only of people rich enough to afford a gravestone.

A3 There may be discussion about which months are in winter.

B Making a frequency table (p 145)

◊ Ask pupils to think of two different ways of tallying from a list.
  • Go through the list, tallying all those in the 0–9 group first; then do the 10–19 group, and so on.
  • Go through the list once only, tallying into the correct group as you go.
Items are less likely to be missed out if the second way is used.

◊ Pupils may already be familiar with grouping tally marks in fives: ||||

124 • 20 *Gravestones*

C Comparing charts (p 146)

◊ This can be done in pairs or as a class discussion.

◊ Two ways of drawing a bar chart are shown. In the first, age is treated as discrete and a gap is left between bars. In the second age is treated as continuous with no gaps between bars, so that a rule is needed for the boundaries.

D Charts galore! (p 148)

Pupils look at good and bad ways of displaying data.

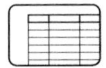

◊ This section will mean more if pupils can use a spreadsheet and draw charts for themselves.

◊ The tilted pie chart can make if difficult to judge the relative sizes of sectors. It loses the empty age groups. It gives no idea of the total sample, but shows the proportions.

The cobweb diagram is hard to read. The shading has no meaning.

The bar chart clearly summarises the data but there is no point in dividing the frequency scale into fifths.

The line graph or frequency polygon retains all the information but can be confusing to interpret. The joining lines have no meaning; they just guide the eye.

The doughnut diagram loses the 10–19 and 20–29 data but shows a gap labelled 30–39!

The 3-D bar chart retains all the information, but the frequency is hard to read.

E Testing a hypothesis (p 150)

Sheet 93

Discussion will be needed to clarify the meaning of the word 'hypothesis'. Emphasise that a hypothesis may turn out to be false.

Challenge (p 150)

John Millington died on 1 September 1694 aged 54.

If he died before his birthday that year then he was born in 1639. If he died after his birthday that year then he was born in 1640.

F Modal group (p 151)

A What gravestones tell us (p 144)

A1 Probably November 1757. Remember that people are not generally buried on the day they die.

A2 Roughly 90 years

A3 It depends on what are the 'winter months'.

J	F	M	A	M	J	J	A	S	O	N	D
1		2	1		2	2	1		1	3	1

B Making a frequency table (p 145)

B1

Age (in years)	Tally	Frequency
0–9	||||	6
10–19		0
20–29		0
30–39		0
40–49			1
50–59	|||	3
60–69	|||	3
70–79			1

The table shows that the people tended to die young or old but not in between.

C Comparing charts (p 146)

C1 There are no missing bars. They have zero height.

C2 70–80

C3 There are no people who died in these age groups.

C4 (a) Pie chart
  (b) Either bar chart
  (c) Either bar chart

E Testing a hypothesis (p 150)

E1 Some pupils will not manage to make a hypothesis of their own. In E2 they can test one of those already suggested.

Possible hypotheses include

'Nobody lived beyond 71.'

'Nobody died aged between 10 and 39.'

'Most people died in November.'

'Boys stood a better chance of surviving childhood.'

E2 'Most deaths occurred in the winter months.'

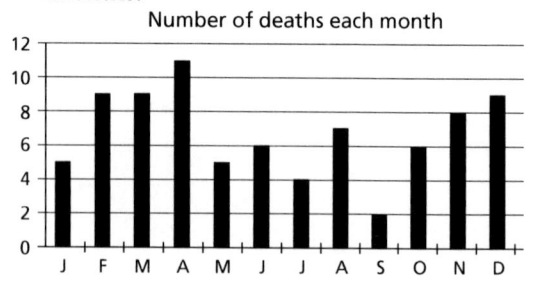

If we define the winter months to be from November to April then the hypothesis seems to be confirmed with 51 dying in that period, 30 outside it.

'Most deaths occurred in the under 10 age group.'

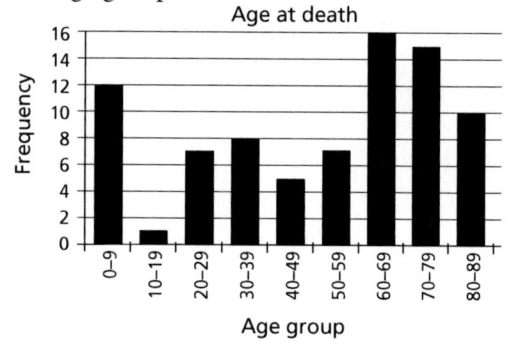

This hypothesis is not confirmed. The bar chart shows that most deaths occurred in the 60–69 age group. It is worth commenting that infant mortality was high.

'If you reached 20, there was a good chance of living to 60.'

Of the 68 people who reached age 20, 41 reached age 60. So the hypothesis is confirmed.

F Modal group (p 151)

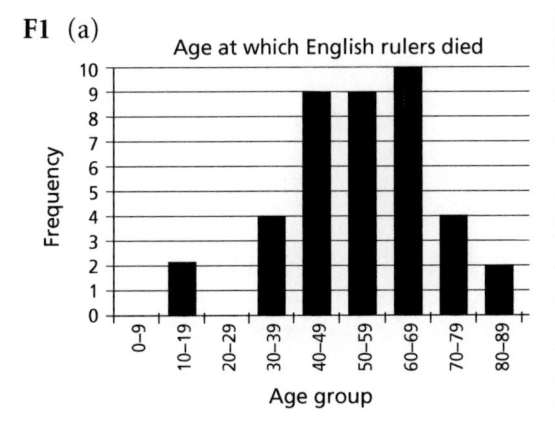

F1 (a)

Age at which English rulers died

[Bar chart showing frequencies by age group: 0–9: 0, 10–19: 2, 20–29: 0, 30–39: 4, 40–49: 9, 50–59: 9, 60–69: 10, 70–79: 4, 80–89: 2]

(b) 60–69

(c) The chart shows that the rulers did not die particularly young. (This could be biased because you might have had to wait for an old parent to die before becoming ruler.) Most managed to get into their forties. (It would be interesting to compare this with the general population.)

Practice booklet

Sections A, B, C and F (p 46)

1 (a) 8   (b) 6   (c) 82

2 (a)

Age at death (years)	Tally	Frequency
0–9	卌 IIII	9
10–19	卌 卌 I	11
20–29	卌 卌 卌 卌 I	21
30–39	卌 III	8
40–49	III	3
50–59	IIII	4
60–69	I	1
70–79	I	1
80–89		0
90–99	I	1

(b) 7

(c) The pupil's grouped frequency chart for the data in (a)

(d) 20–29

(e) The pupil's two sentences, such as:

About two-thirds of the females died before 30 years of age.

Male deaths were more evenly spread.

About half the males died before 30 years of age.

A much larger proportion of the males lived beyond 60.

20 Gravestones • 127

21 Oral questions: recipes (p 153) 7T/41

This page is for teacher-led oral work focusing on simple mental multiplication and division of food quantities.

◊ You could begin by asking some simple questions to get pupils looking at the recipes, for example: 'How much sugar do you need for the apple crumble?', 'Which recipes need butter?'

Typical questioning can begin with 'Suppose you are going to cook sheek kebabs for 8 people …' then (going round the class) '… how much mint would you need?', '… how much minced beef would you need?'

You can then ask about the amount of each ingredient required for these (roughly in order of increasing difficulty) and others of your own.

Topping for 2 pizzas, 3 pizzas, …

20 scones, 30 scones, …

16 pancakes, 24 pancakes

2 jam sandwich cakes

Sheek kebabs for 2 people, 1 person

Apple crumble to serve 3 people, 2 people

Pumpkin soup for 3 people

You could ask diet-conscious questions like 'How much butter is there in one scone?', 'How much cream is there in one serving of pumpkin soup?', 'If the jam sandwich cake serves 5 people, how much sugar is there in each serving?'

Harder questions include those like 'If you had 1 kilogram of plain flour, roughly how many pancakes could you make if you had plenty of the other ingredients?'

22 Work to rule

7T/24, 7C/14

The emphasis in this unit is on finding rules by analysing tile designs. Some pupils may find the concrete experience of making the patterns with tiles or multilink helpful.

Knowledge of the convention that multiplication and division have priority over addition and subtraction is not assumed: brackets are used throughout.

p 154 **A** Mobiles	Finding and using a rule to calculate the number of white tiles given the number of red tiles
p 156 **B** Explaining	Considering explanations of how rules were found
p 158 **C** More designs	Finding and using a rule that uses one or two operations and describing it in words
p 160 **D** Shorthand	Using algebraic shorthand: $w = 2r + 1$
p 162 **E** Snails	Considering a design that gives a simple quadratic rule

Optional
Tiles or multilink in two colours
Squared paper

Practice booklet pages 48 to 51

A Mobiles (p 154)

In this section pupils should become aware that there are two types of rule for these patterns:
- by tabulating results in order we can see how the sequence of white tiles continues – it goes up in 2s
- by looking at the structure of the pieces we can see that the number of whites is equal to double the number of reds plus 3

Pupils should begin to appreciate the advantages of the latter rule.

Optional: Tiles or multilink in two colours, squared paper

◊ You can start by discussing the design of the pieces in the mobile on page 154, perhaps using tiles or multilink.

Tabulate results in order up to, say, 8 red tiles, and look at the pattern in the table. Many pupils will spot that the number of white tiles goes up in 2s, but they should think about why it will continue in 2s.

Ask pupils to imagine the piece with, say, 100 red tiles and how we could calculate the number of white tiles. Pupils should appreciate it would take a long time to continue to add on 2s.

A discussion of the structure of the designs should lead to

- the piece with 100 red tiles has (100 × 2) + 3 whites

and to the general rule

- to find the number of whites, multiply the number of reds by 2 and add 3

All but less able pupils should be able to grasp the shorthand version

number of whites = (number of reds × 2) + 3

Use the discussion to bring out how diagrams can be useful in making the structure of the designs clear, for example:

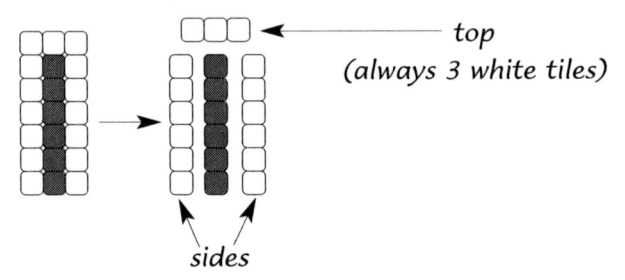

top (always 3 white tiles)

sides

> 'The pupils were very comfortable with the work. I was very pleased with their explanations.'

◊ When asked to explain how to find the number of white tiles for a given number of reds, some pupils may use repeated addition. Encourage all pupils to think about a rule to calculate directly the number of whites in some way. Some pupils will be able to see from the table that because the number of whites 'goes up in 2s' we must multiply the number of reds by 2 when seeking a direct rule.

A5(b) Some pupils may give 102 white tiles as their answer (just adding 2). Others may try to continue the sequence 3, 5, 7, 9 to the 100th term. Encourage them to visualise the piece with 100 reds to enable them to see that the result can be found by multiplying by 2 and adding 1.

A8 Pupils who choose 'number of white tiles = number of red tiles + 2' may be confusing the rule to continue the sequence with the rule to calculate the number of whites given the number of reds.

B **Explaining** (p 156)

◊ This section may be used as the basis for small-group or whole-class discussion. The explanations are not intended to serve as exemplars: encourage pupils to explain in their own way.

C More designs (p 158)

'Went well – but they didn't think to draw diagrams themselves for the table in C3.'

C5 Pupils may have counted on in 1s to answer question C4. Ask them to visualise the piece with 50 reds to help them see that the result can be found by adding 4 to the number of red tiles.

C9(b) Encourage pupils to draw diagrams to illustrate their explanations. Some pupils will find this difficult and should be encouraged to offer their own explanations, however tentative. An oral explanation would be perfectly acceptable at this stage.

C17 Pupils may produce equivalent rules, for example:
number of white tiles = ((number of red tiles + 2) × 2) + 2
number of white tiles = number of red tiles + number of red tiles + 6

This is an opportunity to show equivalence with the rule
number of white tiles = (number of red tiles × 2) + 6

D Shorthand (p 160)

◊ Letters for variables are introduced at this point. Some pupils may continue to feel more confident with rules in the form $w = (r \times 2) + 1$ rather than $w = 2r + 1$. Both forms are acceptable at this stage.

◊ Encourage all pupils, especially the more able, to give written explanations, although oral explanations are acceptable at this stage.

D8 This is intended to provide pupils with an opportunity to structure their own work to find rules. Encourage them to use strategies such as counting tiles in the examples given, drawing more diagrams and tabulating results to help them analyse the diagrams to find a rule.

Pupils could work in groups and present their ideas on one or more of these designs to the whole class.

E Snails (p 162)

E8(b) Pupils are likely to find this difficult.

E9(b) Some pupils may choose to add 13 to the result for 6 red tiles; others may calculate (7 × 7) + 1. This gives an opportunity to compare these two methods.

A Mobiles (p 154)

A1 5 white tiles

A2 (a) The pupil's piece with 5 reds
(b) 11 white tiles

A3

Number of red tiles	1	2	3	4	5	6
Number of white tiles	3	5	7	9	11	13

A4 (a) As the number of red tiles goes up by 1, the number of white tiles goes up by 2 each time.
(b) The pupil's explanations: for example,
an increase of 1 red tile means an extra 2 white tiles, one on each side.

A5 (a) 21 white tiles (b) 201 white tiles

A6 The pupil's method: for example, to find the number of white tiles, multiply the number of red tiles by 2 and add 1; or a method that involves repeated addition of 2.

A7 301 white tiles

A8 number of white tiles = (number of red tiles × 2) + 1

A9 145 white tiles

A10 40 red tiles

A11 The pupil's explanation: for example, the number of white tiles is always odd.

B Explaining (p 156)

B1 (a) The pupil's response
(b) The pupil's response

B2 The pupil's response

B3 The pupil's response and explanation

C More designs (p 158)

C1 10 white tiles

C2 The pupil's bridge with 4 reds; 8 white tiles

C3

Number of red tiles	1	2	3	4	5	6
Number of white tiles	5	6	7	8	9	10

C4 (a) 14 white tiles
(b) The pupil's bridge with 10 reds

C5 54 white tiles

C6 1004 white tiles

C7 The pupil's method: for example, add 4 to the number of red tiles.

C8 40 white tiles

C9 (a) number of white tiles = number of red tiles + 4
(b) The pupil's explanation

C10 100 white tiles

C11 (a) 48 red tiles
(b) The pupil's method: for example, 52 − 4 = 48

C12 (a) The pupil's surround with 6 reds
(b) 18 white tiles

C13

Number of red tiles	1	2	3	4	5	6
Number of white tiles	8	10	12	14	16	18

C14 26 white tiles

C15 206 white tiles

C16 The pupil's explanation

C17 (a) Any rule equivalent to
number of white tiles = (number of red tiles × 2) + 6
(b) The pupil's explanation
(c) 174 white tiles

C18 60 red tiles

C19 The pupil's explanation: for example, the number of white tiles is always even.

D Shorthand (p 160)

D1 A and F, B and H, C and E, D and G

D2 Any letters can be used to stand for the *number of white tiles* and the *number of red tiles*. We have used *w* for the *number of white tiles* and *r* for the *number of red tiles*.

(a) $w = 4r + 3$ or equivalent

(b) $w = 5r$ or equivalent

(c) $w = 2r + 7$ or equivalent

(d) $w = r + 6$

(e) $w = 3r - 1$ or equivalent

D3 23 white tiles

D4 243 white tiles

D5 (a) Any rule equivalent to
number of white tiles =
(number of red tiles × 4) + 3

(b) $w = 4r + 3$ or equivalent

(c) The pupil's explanation of why the rule works

D6 403 white tiles

D7 (a) The pupil's design that fits $w = 2r + 2$

(b) The pupil's explanation (with diagrams)

D8 For each rule (or its equivalent) the pupil should give an explanation of why it works.
Set A: $w = r + 1$
Set B: $w = r + 4$
Set C: $w = 3r + 1$

E Snails (p 162)

E1 10 white tiles

E2 (a) The pupil's snail with 4 red tiles

(b) 17 white tiles

E3 37 white tiles

E4 101 white tiles

E5 401 white tiles

E6 The pupil's explanation

E7

Number of red tiles	1	2	3	4	5	6
Number of white tiles	2	5	10	17	26	37

E8 (a) The pupil's description: for example, the number of white tiles goes up by 3, then 5, then 7, then 9, … (these numbers go up by 2 each time).

(b) The pupil's explanation: for example, each snail is made by adding an L-shape of tiles to the previous one and the L-shape gets bigger by 2 each time.

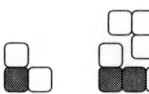

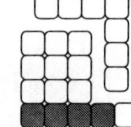

E9 (a) 50 white tiles

(b) The pupil's method: for example, $37 + 13 = 50$, $(7 \times 7) + 1 = 50$

E10 12 red tiles

E11 No; the pupil's explanation

What progress have you made? (p 163)

1 17 white tiles

2 302 white tiles

3 The pupil's explanation of the rule $w = 3r + 2$

4 (a) $w = r + 8$

(b) $w = 4r + 1$ or $w = (r \times 4) + 1$

5 (a) $w = 2r + 4$ or equivalent

(b) The pupil's explanation

(c) 104 white tiles

22 Work to rule • 133

Practice booklet

Sections A, B and C (p 48)

1 3 whole circles

2 7 whole circles

3 (a) The pupil's frieze with 5 tiles
   (b) 9 whole circles

4

Number of tiles	1	2	3	4	5	6
Number of circles	1	3	5	7	9	11

5 (a) The number of circles goes up by 2 each time a tile is added.
   (b) The pupil's explanation: for example, each time a tile is added one half circle on the join is completed and one circle in the centre is added.

6 (a) 15   (b) 19

7 199

8 The pupil's explanation: for example, double the number of tiles and subtract 1.

9 299

10 (a) number of circles = (number of tiles × 2) − 1
   (b) The pupil's explanation

11 (a) 53   (b) 161   (c) 399

12 (a) 13
   (b) The pupil's explanation

13 45

14 The pupil's explanation: for example, the number of circles is always an odd number.

Section D (p 49)

1 A and C, B and D, E and F

2 (a) $c = t + 7$ (b) $c = 5t - 3$

3 16

4 (a) The pupil's frieze with 5 tiles
   (b) 13

5 19

6 (a) 28   (b) 58   (c) 298

7 The pupil's explanation: for example, multiply the number of tiles by 3 and subtract 2.

8 (a) $c = 3t - 2$
   (b) The pupil's explanation

9 (a) 259   (b) 283   (c) 2998

10 (a) $c = 3t - 1$
   (b) The pupil's explanation
   (c) (i) 71   (ii) 149
       (iii) 299

*11 (a) $c = \dfrac{3t}{2} - 1$
   (b) The pupil's explanation

23 Triangles

7C/7

As this is a long unit, you may wish to break it into two parts, for example after section D.

p 164	**A** Drawing a triangle accurately	Constructing a triangle given three sides, using compasses
p 166	**B** Equilateral triangles	Making a net for an octahedron
p 168	**C** Isosceles triangles	
p 170	**D** Scalene triangles	
p 171	**E** Using angles	Constructing a triangle given sides and angles
p 174	**F** Angles of a triangle	
p 176	**G** Using angles in isosceles triangles	

Essential	**Optional**
Plain paper	Compasses for board or OHP
Compasses	Triangular dotty paper
Scissors	
Tracing paper	
An envelope to keep cut or traced triangles	
Thin card (for nets)	
Glue	
Angle measurer	
Sheet 88	
Practice booklet pages 52 to 54	

A Drawing a triangle accurately (p 164)

> Plain paper, compasses, tracing paper or scissors

◊ A good way to begin is by sketching an 8, 5, 10 triangle on the board and challenging the class to draw it accurately with pencil and ruler only. They may get an accurate result, but probably only after some trial and error. This should help them see the advantage of using compasses.

◊ If there are discrepancies when pupils compare their triangles with their neighbours', the problems should be sorted out and the triangles drawn again. They will be needed for later questions.

Investigation

For three lengths to make a triangle, the longest must be less than the sum of the other two.

B Equilateral triangles (p 166)

> The triangles made in section A, compasses, scissors, thin card, glue, (possibly) triangular dotty paper

◊ Pupils who, because of poor manipulative skills, are likely to be discouraged rather than helped by doing so much work with compasses could work on triangular dotty paper.

◊ Pupils could design and make other polyhedra with equilateral triangles as faces. Here is an easy-to-make net for a regular icosahedron.

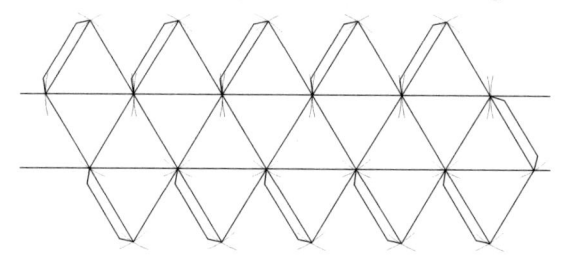

C Isosceles triangles (p 168)

> The triangles made in section A, sheet 88 (preferably on thin card), compasses, scissors, glue

◊ The fact that an equilateral triangle is a special case of an isosceles triangle may come up in the answers to questions and in discussion. There is no need to make a big thing of it at this stage.

C5 You may need to give help on naming triangles by the letters of their vertices.

D Scalene triangles (p 170)

> The triangles made in section A

◊ Ensure pupils realise that a right-angled triangle can be scalene.

E Using angles (p 171)

This section includes drawing a triangle given one side and two angles, and given two sides and the angle between them.

> Angle measurer

F Angles of a triangle (p 174)

> Angle measurer, scissors

◊ Ask everybody to draw a triangle, measure its angles and add them together. Enough of the results should be close to 180° to blame the discrepancy on inaccurate drawing and measurement!

◊ For the torn-off angles demonstration you may need to remind pupils about angles on a line.

Of course, neither of these approaches amounts to a proof. The proof is taken up in later material.

G Using angles in isosceles triangles (p 176)

◊ The 'Can you make it?' box can be discussed with pupils when you draw together the work in the unit.

B Equilateral triangles (p 166)

B1 (a) Triangles C and F
   (b) They are all 60°.

B2 Yes, you can fold it so two sides and two angles match up with one another. The fold line is the mirror line.

B3 6 vertices

B4 12 edges

B5 All the angles and sides of the triangular 'faces' are the same.

C Isosceles triangles (p 168)

C1 Triangles B, C, E and F (C and F are also equilateral).

C2 (a) It is 90° (a right angle).
   (b) They are equal.
   (c) It has reflection symmetry, with the fold line as a line of symmetry.

C3 The pupil's model

C4 Between 5 and 7 cm is a reasonable result (set squares are useful measuring aids). The important thing is that pupils don't just measure the sloping edge length or even the distance from the midpoint of the side of the square to the top of the vertex (a result of about 7.5 cm would indicate this).

23 Triangles • 137

C5 These triangles are isosceles. Finding just some of them is a fair achievement.

ACO, ECN, ACN, NCO, OCE, BOC, DNC, ANO, EON, FON, GNO, KGO, MFN, NIO, NLO, BCF, CDG, CFO, CGN, FIN, GIO

Diagonal cut puzzle

The triangles cut out are two pairs of isosceles triangles.

These are the two ways of fitting them together.

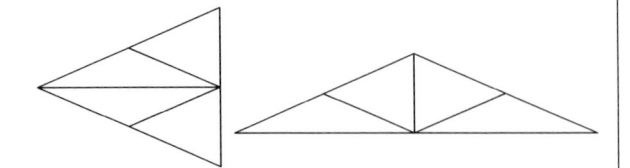

D Scalene triangles (p 170)

D1 Triangles A and D

D2 ABC, ABD, ACE, ADE

D3 (a) Isosceles
(b) Scalene
(c) Equilateral
(d) Isosceles
(e) Scalene
(f) Isosceles
(g) Equilateral

D4 No. Pupils can think about what would happen if they tried folding so that one side went on to another. They would be different lengths so they would not match.

E Using angles (p 171)

E1 The pupil's triangle

E2 The pupil's triangles

E3 It is not possible to complete the triangle because two of the lines are parallel.

E4 The pupil's triangle

E5 The pupil's triangles

E6 (a) The pupil's triangle
(b) BC = 7.5 cm, angle at B = 49°, angle at C = 61°

E7 (a) The pupil's triangle
(b) XZ = 7.4 cm, YZ = 12.4 cm, angle at Z = 20°

E8 (a) The pupil's triangle
(b) AC = 12.5 cm, angle at A = 61°, angle at C = 34°

E9 (a) The pupil's triangle
(b) QR = 12.5 cm, angle at Q = 32°, angle at R = 23°

E10 It is impossible to draw a triangle (because the circle centre B with radius 6 cm does not cross the line through A at 40° to AB).

E11 Two different triangles can be drawn (because the circle centre Q with radius 7 cm crosses the line through P at two points).

F Angles of a triangle (p 174)

F1 (a) 50° (b) 60° (c) 55°
(d) 80° (e) 20°

F2 (a) 75° (b) 121° (c) 85°
(d) 66° (e) 91°

F3 Each angle is 60°, because 180 ÷ 3 = 60.

F4 (a) 60° (b) 25° (c) 17°
(d) 62° (e) 33°

F5 $a = 60°$ $b = 120°$ $c = 70°$
$d = 110°$ $e = 80°$

F6 $a = 70°$ $b = 65°$ $c = 128°$
$d = 80°$ $e = 100°$ $f = 148°$

F7 To draw the triangle, the third angle (40°) must be worked out first.

Ⓖ Using angles in isosceles triangles (p 176)

G1 They are equal.

G2 $a = 75°$ $b = 30°$ $c = 24°$
     $d = 132°$ $e = 54°$ $f = 72°$
     $g = 18°$ $h = 144°$ $i = 77°$
     $j = 26°$

G3 $a = 70°$ $b = 70°$ $c = 40°$
     $d = 40°$ $e = 66°$ $f = 66°$
     $g = 30°$ $h = 30°$ $i = 48°$
     $j = 48°$

G4 (a) Either 40° and 100° or 70° and 70°
     (b) Either 72° and 36° or 54° and 54°
     (c) 25° and 25°

G5 The triangles will not fit together because 50° does not go into 360° exactly.

G6 9 sides

G7 30°

G8 144°

What progress have you made? (p 178)

1. The pupil's triangle

2. (a) ABD
     (b) EBD, ABD
     (c) ABE, ADE, BEC, DEC, ABC, ADC
     (d) ABC, EBC, ADC, EDC

3. The pupil's triangles

4. $a = 112°$ $b = 57°$

5. $a = 67°$ $b = 46°$ $c = 38°$

Practice booklet

Sections A, B, C and D (p 52)

1. The pupil's triangle
   The angles should be 87°, 51°, 42°.

2. The pupil's triangle
   The angles should be 90°, 53°, 37°.

3. (a) C, D   (b) A, E (also C and D)
    (c) B, F, G   (d) B, F

Section E (p 53)

1. (a) The pupil's triangle; the third angle should be 105°.
    (b) The pupil's triangle; the third angle should be 80°.

2. (a) The pupil's triangle; the third side should be 4.5 cm.
    (b) The pupil's triangle; the third side should be 11.5 cm.

Sections F and G (p 53)

1. $a = 60°$ $b = 40°$ $c = 25°$

2. $a = 65°$ $b = 115°$ $c = 49°$ $d = 131°$
   $e = 38°$ $f = 67°$ $g = 43°$

3. $a = 74°$ $b = 74°$ $c = 55°$ $d = 70°$
   $e = 22°$ $f = 22°$ $g = 40°$

4. Yes, its angles are 90°, 45,° 45°.

5. Either 50° and 80° or 65° and 65°

6. (a) 45°   (b) $67\frac{1}{2}°$

24 Decimal calculation

7T/29, 7C/15

This unit covers adding and subtracting decimals up to two decimal places and multiplying and dividing them by single-digit numbers. All the work is to be done without the use of a calculator.

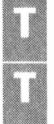

p 179 **A** Adding and subtracting – one decimal place

p 181 **B** Adding and subtracting – up to two decimal places

p 183 **C** Multiplying by a single-digit number

p 184 **D** Dividing by a single-digit number

p 186 **E** Mixed problems

Essential	**Optional**
Sheet 44	Sheet 125
Practice booklet pages 55 to 58	

A Adding and subtracting – one decimal place (p 179)

Sheet 44

◊ The teacher-led introduction should include discussion of both mental and written methods. The numbers in the loop can be used for this and some examples of suitable questions (which you can supplement with your own) are shown beside the loop. You could ask pupils to make up some questions of their own for other pupils to answer, from these numbers or from further loops.

Include mental work such as
- 12.4 + 0.1
- 0.8 + 0.2 (decimal complements in 1)
- 1.3 – 0.9
- 3 – 0.9

◊ Include work on standard column procedures for adding and subtracting. Pupils often find using the standard column procedure tricky for examples such as 12 – 4.3: you may need to remind them that 12 can be written as 12.0.

140 • 24 *Decimal calculation*

◊ A suggestion for a game for two players called 'One or two' (not shown in the pupil's book) is given below.

One or two

Sheet 125 (one set of cards per pair)

◊ Each pair puts a set of cards from sheet 125 face up on the table. Players take turns to pick up a card.

The first player to have in their hand **three** cards that add up to 1 or 2 wins the round. They get one point.

The first person to win 10 points wins the game.

B Adding and subtracting – up to two decimal places (p 181)

◊ As before, the teacher-led introduction should include discussion of mental and written methods. The weights of cheese can be used for this and some examples of suitable questions are shown below.
- How much do the two pieces of cheese weigh in total?
- How much heavier is the Cheddar than the Brie?

Include mental work such as
- $1.35 - 0.6$
- $23.78 + 0.01$
- $169.31 - 0.1$

◊ As before, include work on standard column procedures. Pupils often find this tricky for examples such as $4.6 - 1.23$: you may need to emphasise the necessity to line up decimal points and remind them that 4.6 can be written as 4.60.

◊ Estimating answers first by rounding figures to the nearest whole number can help pupils identify errors and this should be encouraged.

C Multiplying by a single-digit number (p 183)

◊ Some teacher input may be required. You may need to discuss how to make sure the decimal point is placed correctly in the answer. Again it is helpful to have a rough idea of how big the answer should be before calculating it.

D Dividing by a single-digit number (p 184)

◊ Include in your discussion more difficult examples such as $1.4 \div 4$ where, to use the standard division algorithm, a zero needs to be added. For $3.27 \div 3$ some will respond with 1.9 rather than 1.09 and careful discussion will be needed here.

E Mixed problems (p 186)

A Adding and subtracting – one decimal place (p 179)

A1 (a) 7.2 (b) 4 (c) 3.9 (d) 4.1
 (e) 11.2

A2 (a) 2.5 (b) 6.1 (c) 1.9 (d) 0.9
 (e) 1.5

A3 (a) Cheddar: 1.2 kg   Stilton: 0.9 kg
 (b) 0.3 kg   (c) 2.1 kg

A4 (a) 2.8 + 4.5 = **7.3**
 (b) **3.4** + **6.3** = 9.7
 (c) 1.6 + **1.5** = 3.**1**
 (d) 23.**3** + **4**.8 = 28.1

A5 6.4 kg

A6 (a) 9.5 + 8.1 = 17.6 or
  9.1 + 8.5 = 17.6
 (b) 5.9 + 1.8 = 7.7 or
  5.8 + 1.9 = 7.7

A7 (a) 36.2 (b) 70.7
 (c) 17 or 17.0 (d) 43.7

A8 (a) 6.9 – 4.5 = **2.4**
 (b) 7.0 – **1.5** = 5.5
 (c) **5.1** – 2.4 = 2.7
 (d) 37.2 – 5.5 = **31.7**

A9 (a) 7.6 – 2.3 = 5.3
 (b) 7.2 – 6.3 = 0.9 or
  3.6 – 2.7 = 0.9

A10 (a) 4.9 (b) 10.2
 (c) 15.9 (d) 41.3

A11 a = 28.3 cm, b = 15.5 cm

A12 (a) 3.0 km (accept 2.9 to 3.2)
 (b) 11.6 km (accept 11.5 to 12)
 (c) 15.3 km (accept 15.0 to 15.8)
  Best route: John – Ann – Joseph – Paul – Mary – Peter – John
 (d) The canal lock is 2 km closer (accept 1.9 to 2.1).

A13 1.7 cm

A14 11.6 km

B Adding and subtracting – up to two decimal places (p 181)

B1 (a) 0.1 (b) 0.3 (c) 0.01

B2 (a) 4.77 (b) 8.22
 (c) 19.7 or 19.70 (d) 39.92

B3 3.73 kg

B4 (a) 5 (b) 18.92 (c) 7.45

B5 (a) B: 19 – 5 (b) 13.72

B6 (a) 3.84 (b) 10.89
 (c) 8.36 (d) 6.09

B7 (a) 2.19 (b) 3.23

B8 A, C and D are wrong.
 A: 2.32 + 4.9 = 7.22
 C: 9.3 – 5.75 = 3.55
 D: 6.54 – 3.9 = 2.64

B9 (a) 118.47 (b) 1.32
 (c) 4.42 (d) 115.64

B10 (a) **5**.34 + 16.76 = 22.10
 (b) 6.71 – 0.23 = **6**.4**8**
 (c) 5.02 – 1.**6**1 = **3**.41
 (d) **1**2**3**.**5**0 – 17.32 = 106.18

B11 1.05 m

B12 Yes, the gap is 0.34 m

B13 32.55 kg

B14 (a) Mike: 4.23 + 2.87 + 2.6 = 9.7 kg
  Justin: 3.14 + 3.6 + 4.71 = 11.45 kg
  Asad: 3.92 + 3.44 + 3.7 = 11.06 kg
 (b) Justin (c) 0.39 kg

B15 1.75 kg, 2.2 kg, 3.05 kg, 2.85 kg
 (with a total weight of 9.85 kg)

C Multiplying by a single-digit number (p 183)

C1 (a) 1.4 (b) 1.5 (c) 0.8
 (d) 1.8 (e) 2 or 2.0

C2 (a) 8.1 litres (b) 7.2 kg
 (c) 5.4 litres

C3 (a) 0.05 (b) 6.03 (c) 0.24
 (d) 0.52 (e) 0.56

C4 (a) 55.8 (b) 16.5 (c) 26.55 (d) 3.72

C5 (a) 9.12 (b) 31.2 (c) 63.18
 (d) 31.22 (e) 2466

C6 49.05 kg

C7 (a) 8.15 m (b) 5.25 litres
 (c) £8.64

C8 (a) 12.36 × 4 = 49.44
 (b) 15.06 × 3 = 45.18
 (c) 0.56 × 9 = 5.04
 (d) 60.92 × 8 = 487.36

D Dividing by a single-digit number (p 184)

D1 (a) 0.5 (b) 2.4 (c) 1.6
 (d) 0.4 (e) 0.4

D2 1.4 metres

D3 0.2 litres

D4 2.4 kg

D5 (a) 2.5 (b) 2.6 (c) 5.4 (d) 2.4
 (e) 2.2 (f) 8.5 (g) 16.4 (h) 0.6

D6 £2.90

D7 £1.40

D8 £1.95

D9 £10.15

D10 (a) £1.42, £1.45
 (b) The larger bag gives you more as the onions are 3p cheaper per kilogram.

D11 (a) 1.34 (b) 4.84 (c) 5.91
 (d) 0.21 (e) 7.22

D12 The pupil's explanation such as 'You need an extra decimal place to deal with the remainder (of 0.1).'

D13 (a) 2.65 (b) 1.35 (c) 2.45
 (d) 3.05 (e) 4.62

D14 (a) 1.83 (b) 2.04 (c) 1.65 (d) 0.82

D15 (a) 3.58 (b) 2.08 (c) 1.452
 (d) 10.06 (e) 3.295

D16 £1.25

D17 (a) 1.5 (b) 2.25 (c) 0.75
 (d) 14.25 (e) 3.625

E Mixed problems (p 186)

E1 22 square metres

E2 (a) 6.19 m (b) 0.03 m
 (c) 15.2 m (d) 2.32 m

E3 3.45 m

E4 (a) The box of 5 gives you more for your money. Each fruit pie in this box costs 21p but in the larger box each pie costs 24p.
 (b) The box of 10 gives you more for your money. Each egg in this box costs 17p but in the smaller box each egg costs 18p.

E5 0.26 m or 26 cm

E6 Yes, the total length of the two pieces he needs is 4.46 m and 4.5 m is longer than this.

E7 16.28 kg

What progress have you made? (p 187)

1 (a) 14.2 (b) 8.9 (c) 4.7 (d) 6.3

2 (a) 2.4 + 3.8 = 6.2
 (b) 3.1 − 1.7 = 1.4

3 (a) 20.85 (b) 19.15 (c) 0.63
  (d) 22.14
4 (a) 2.5 (b) 2.1 (c) 10.8
  (d) 3.75 (e) 21.42 (f) 192.06
5 (a) 0.3 (b) 1.7 (c) 2.46
  (d) 2.48 (e) 2.06 (f) 3.245
6 (a) 2.7 litres (b) 12.25 litres
  (c) £1.95 (d) 3.25 m
  (e) 0.45 m

Practice booklet

Sections A and B (p 55)

1 (a) 6.8 (b) 5.7 (c) 10
  (d) 2.6 (e) 5.8 (f) 2.1
  (g) 3.2 (h) 0.5
2 (a) 0.6 kg (b) 4.4 kg
3 (a) 41.2 (b) 42.9 (c) 51.3
  (d) 40.1 (e) 16.1 (f) 22.8
  (g) 14.5 (h) 20.6 (i) 105.4
4 (a) 12.4 cm (b) 2.8 cm
5 (a) Andy 4.41 kg, Fi 4.58 kg
  (b) Fi, by 0.17 kg
6 (a) 11.13 (b) 6.25 (c) 15.73
  (d) 21.26 (e) 43.09 (f) 204.21
7 (a) 1.73 (b) 1.58 (c) 180.55
  (d) 1.24 (e) 23.86 (f) 246.97
  (g) 1.04 (h) 7.49 (i) 331.03

Section C (p 56)

1 4.2 kg
2 (a) 3.5 (b) 1.2 (c) 1.8
  (d) 3.5 (e) 4
3 20.4 m
4 (a) 12.6 (b) 22 (c) 84.6
  (d) 26.1 (e) 5.08 (f) 44.8
  (g) 147.63 (h) 69.84
5 240.45 kg

Section D (p 57)

1 3.9 cm
2 £1.39
3 (a) 0.7 (b) 1.71 (c) 2.7
  (d) 12.82 (e) 37.79 (f) 1.68
  (g) 2.04 (h) 6.14 (i) 13.61
4 0.85 m
5 (a) 1.45 (b) 1.15 (c) 3.26
  (d) 0.76 (e) 14.04 (f) 10.09
  (g) 2.185 (h) 1.205
6 (a) 2.5 (b) 1.4 (c) 4.75
  (d) 2.125

Section E (p 58)

1 0.3 litres
2 6.3 metres
3 1.56 metres
4 8.75 kg
5 0.06 metres or 6 cm
6 Yes, the boxes weigh 369.6 kg in total.
7 0.25 litres or 250 ml
8 £2.37
9 No, the total length she needs is 8.07 m.
10 (a) £1.42 (b) 6p left

25 Equivalent fractions and ratio   7C/8

p 188 **A** Equivalent fractions and simplifying

p 191 **B** Comparing, adding and subtracting

p 192 **C** Mixed numbers

p 193 **D** Comparing parts

Practice booklet pages 59 to 61

A Equivalent fractions and simplifying (p 188)

◊ 'Pie' diagrams can be used to explain why the numerator and denominator are both multiplied by the same number. For example, in the case of $\frac{3}{4}$, each of the quarters can be subdivided into, say, 5 equal parts, giving $\frac{15}{20}$ as an equivalent fraction.

You may need to emphasise that equivalence works both ways: $\frac{3}{6}$ is equivalent to $\frac{1}{2}$ and vice versa.

◊ Some pupils may produce a list of equivalent fractions by doubling the numerator and denominator each time, for example:

$$\frac{1}{3} = \frac{2}{6} = \frac{4}{12} = \frac{8}{24} = \ldots$$

Emphasise that this strategy leads to missed fractions, for example $\frac{3}{9}$.

B Comparing, adding and subtracting (p 191)

◊ Here, to build confidence, a common denominator is given, and pupils compare, add and subtract fractions that can be expressed in terms of it. Later, in *Book 8S*, they are given fractions and have to find the common denominator themselves.

C Mixed numbers (p 192)

25 Equivalent fractions and ratio • 145

D Comparing parts (p 193)

◊ This section highlights the difference between using a fraction to say what proportion something is of a whole and using the idea of ratio to compare the relative sizes of the parts. For example, in question D1 one-third of the members are girls, but the ratio of girls to boys is 1:2.

The use of ratio is very informal here, so you can decide whether to use the colon notation for ratio at this stage. There is a full treatment of ratio in *Book 8S*.

D10 Some pupils may give '8 to 4' as an answer, in which case you can ask whether they can give the ratio in a simpler way.

A Equivalent fractions and simplifying (p 188)

A1 $\frac{3}{12}$

A2 $\frac{2}{10}$

A3 $\frac{6}{8}, \frac{9}{12}, \frac{12}{16}$

A4 $\frac{3}{4} = \frac{6}{8}$

A5 $\frac{2}{3} = \frac{4}{6}$

A6 (a) $\frac{1}{2} = \frac{2}{4}$ (b) $\frac{1}{3} = \frac{3}{9}$ (c) $\frac{5}{6} = \frac{10}{12}$
(d) $\frac{3}{5} = \frac{9}{15}$ (e) $\frac{4}{5} = \frac{16}{20}$

A7 (a) $\frac{1}{2}$ (b) $\frac{3}{4}$ (c) $\frac{1}{4}$ (d) $\frac{3}{4}$ (e) $\frac{2}{3}$

A8 $\frac{1}{4}$

A9 (a) $\frac{1}{4}$ (b) $\frac{3}{8}$ (c) $\frac{2}{3}$ (d) $\frac{3}{7}$ (e) $\frac{1}{5}$

A10 $\frac{1}{3}$

A11 $\frac{1}{3}$

A12 $\frac{3}{4}$

A13 (a) $\frac{2}{5}$ (b) $\frac{3}{8}$ (c) $\frac{2}{5}$

A14 $\frac{3}{5}$

A15 (a) $\frac{5}{8}$ (b) $\frac{1}{4}$ (c) $\frac{1}{8}$

A16 $\frac{3}{7}$

A17 (a) $\frac{2}{3}$ (b) $\frac{5}{8}$ (c) $\frac{2}{5}$
(d) Cannot be simplified   (e) $\frac{2}{5}$

A18 (a) $\frac{1}{3}$ (b) Cannot be simplified   (c) $\frac{2}{3}$
(d) Cannot be simplified   (e) $\frac{2}{5}$ (f) $\frac{2}{3}$
(g) $\frac{3}{7}$ (h) Cannot be simplified   (i) $\frac{5}{8}$
(j) $\frac{4}{15}$

A19 $\frac{2}{3}$

Fraction hunt

Proper fractions: $\frac{1}{2}$ $\frac{1}{3}$ $\frac{1}{4}$ $\frac{1}{6}$ $\frac{2}{3}$ $\frac{3}{4}$

Improper fractions: $\frac{2}{1}$ $\frac{3}{1}$ $\frac{4}{1}$ $\frac{6}{1}$ $\frac{3}{2}$ $\frac{4}{3}$

(and $\frac{1}{1}$ if a digit can be repeated)

B Comparing, adding and subtracting (p 191)

B1 (a) (i) $\frac{8}{12}$ (ii) $\frac{9}{12}$ (b) $\frac{1}{2}$ $\frac{7}{12}$ $\frac{2}{3}$ $\frac{3}{4}$

B2 (a) $\frac{1}{4}$ (b) $\frac{1}{3}$ (c) $\frac{2}{3}$ (d) $\frac{7}{12}$ (e) $\frac{5}{12}$

B3 (a) (i) $\frac{16}{20}$ (ii) $\frac{18}{20}$ (b) $\frac{3}{4}$ $\frac{4}{5}$ $\frac{17}{20}$ $\frac{9}{10}$

B4 (a) $\frac{3}{10}$ (b) $\frac{1}{2}$ (c) $\frac{1}{5}$ (d) $\frac{1}{4}$ (e) $\frac{1}{4}$
(f) $\frac{3}{5}$ (g) $\frac{9}{20}$ (h) $\frac{1}{20}$ (i) $\frac{9}{20}$ (j) $\frac{7}{10}$

C Mixed numbers (p 192)

C1 There are 4 quarters in 1.
So there are 8 quarters in 2.
The extra quarter makes 9 quarters.

C2 (a) 16   (b) 23   (c) 51   (d) 44

C3 (a) $1\frac{2}{5}$ (b) $2\frac{3}{5}$ (c) $1\frac{1}{4}$ (d) $2\frac{3}{4}$ (e) $3\frac{1}{6}$

C4 (a) $1\frac{3}{5}$ (b) $1\frac{1}{3}$ (c) $1\frac{1}{5}$ (d) $2\frac{1}{4}$ (e) $2\frac{1}{10}$

C5 (a) $5 \times \frac{3}{4} = \frac{15}{4}$ (b) $4 \times \frac{4}{5} = \frac{16}{5}$
(c) $9 \times \frac{7}{10} = \frac{63}{10}$ (d) $5 \times \frac{5}{8} = \frac{25}{8}$

C6 (a) $3\frac{3}{4}$ (b) $3\frac{1}{5}$ (c) $6\frac{3}{10}$ (d) $3\frac{1}{8}$

C7 (a) $2\frac{2}{5}$ (b) $7\frac{1}{3}$ (c) $4\frac{4}{5}$ (d) $2\frac{5}{8}$ (e) $8\frac{1}{10}$

D Comparing parts (p 193)

D1 (a) $\frac{1}{3}$ (b) $\frac{2}{3}$

D2 (a) $\frac{1}{4}$
(b) There are **three** times as many women in the pool as men.

D3 (a) $\frac{4}{5}$
(b) She won **four** times as many matches as she lost.

D4 (a) $\frac{1}{6}$ (b) $\frac{5}{6}$ (c) 7 (d) 35

D5 There are **three** times as many red beads as green beads.

D6 (a) 40 (b) 160

D7 $\frac{2}{5}$

D8 (a) 28 (b) 35

D9 (a) 3 to 1 (b) 4 (c) 16

D10 (a) 2 to 1 (b) $\frac{2}{3}$

D11 (a) 12 (b) 20

*__D12__ (a) 5 (b) 25

What progress have you made? (p 194)

1 $\frac{3}{5}$

2 $\frac{16}{30}$ (or $\frac{8}{15}$)

3 27

Practice booklet

Section A (p 59)

1 The pupil's fractions equivalent to $\frac{4}{6}$: for example $\frac{2}{3}, \frac{8}{12}$

2 The pupil's fraction equivalent to $\frac{5}{6}$: for example $\frac{10}{12}$

3 (a) $\frac{1}{2} = \frac{5}{10}$ (b) $\frac{1}{5} = \frac{4}{20}$
(c) $\frac{10}{20} = \frac{5}{6}$ (d) $\frac{6}{18} = \frac{2}{6}$

4 (a) The pupil's two fractions equivalent to $\frac{1}{4}$: for example $\frac{2}{8}, \frac{3}{12}$
(b) The pupil's two fractions equivalent to $\frac{6}{8}$: for example $\frac{3}{4}, \frac{12}{16}$
(c) The pupil's two fractions equivalent to $\frac{5}{8}$: for example $\frac{10}{16}, \frac{15}{24}$
(d) The pupil's two fractions equivalent to $\frac{8}{10}$: for example $\frac{4}{5}, \frac{16}{20}$

5 (a) $\frac{1}{4}$ (b) $\frac{1}{4}$ (c) $\frac{1}{5}$ (d) $\frac{1}{6}$
(e) $\frac{4}{5}$ (f) $\frac{2}{3}$ (g) $\frac{3}{4}$ (h) $\frac{2}{3}$
(i) $\frac{5}{6}$ (j) $\frac{5}{6}$ (k) $\frac{5}{8}$ (l) $\frac{3}{4}$

6 (a) $\frac{2}{3}$ (b) $\frac{5}{8}$

7 $\frac{5}{9}$

Sections B and C (p 60)

1 (a) (i) $\frac{12}{18}$ (ii) $\frac{8}{18}$ (b) $\frac{1}{3}, \frac{7}{18}, \frac{1}{2}, \frac{5}{9}$
(c) (i) $\frac{1}{6}$ (ii) $\frac{7}{18}$ (iii) $\frac{13}{18}$ (iv) $\frac{1}{18}$

2 (a) 9 (b) 15 (c) 21 (d) 35

3 (a) $2\frac{3}{4}$ (b) $5\frac{1}{3}$ (c) $6\frac{1}{5}$ (d) $3\frac{5}{6}$

4 (a) $1\frac{1}{5}$ (b) $2\frac{3}{6}$ or $2\frac{1}{2}$ (c) $1\frac{8}{10}$ or $1\frac{4}{5}$
(d) $1\frac{4}{9}$ (e) 1 (f) $1\frac{5}{18}$

5 5

6 (a) $1\frac{4}{5}$ (b) $1\frac{8}{10}$ or $1\frac{4}{5}$ (c) $1\frac{7}{8}$ (d) $1\frac{5}{9}$

7 (a) $\frac{4}{5} \times 3 = 2\frac{2}{5}$ (b) $\frac{3}{4} \times 7 = 5\frac{1}{4}$
(c) $\frac{4}{9} \times 5 = 2\frac{2}{9}$

Section D (p 61)

1 $\frac{2}{3}$

2 (a) $\frac{3}{4}$ (b) 3
(c) There are **three** times as many purple beads as blue beads.

3 18

4 (a) 1 to 4 (b) $\frac{4}{5}$

25 Equivalent fractions and ratio • 147

26 Area and perimeter 7T/13, 7C/17

Most of the work involves rectangles. There is introductory work on right-angled triangles.

p 195	**A** Exploring perimeters	
p 196	**B** Rectangles	Area and perimeter of rectangles (sides in whole centimetres)
p 198	**C** Square metres	Area and perimeter of rectilinear compound shapes (sides in whole metres)
p 201	**D** Bringing in triangles	Area of right-angled triangles
p 202	**E** Using decimals	Area of rectangles where sides are a multiple of 0.5

Essential
Centimetre squared paper

Practice booklet pages 61 to 64

Optional
Newspapers

A Exploring perimeters (p 195)

◊ These investigations are ordered by in increasing difficulty. Pupils can go as far with them as they can, but all should become familiar with the word *perimeter* and should come to see that a particular number of squares (producing shapes of a fixed *area*) can give rise to different perimeters. The perimeters are all even numbers of centimetres (investigation 3) because before the squares are put together their total perimeter is an even number of centimetres (actually a multiple of 4) and each time two edges are put together the total perimeter falls by 2 centimetres, remaining even.

The chart on the next page shows the possible perimeters for investigations 2, 4 and 5.

The maximum perimeter for a given number of squares n is $2n + 2$ or $2(n + 1)$ and arises when all the squares are in a straight line or form a shape 'one square wide' with bends in it (the least 'compact' arrangement). Some pupils should be able to explain, in their own words, why such a relationship applies. This linear relationship is shown by the upper right-hand edge of the block of ticks on the chart.

> 'Students discussed the compactness of the shapes.'

148 • 26 Area and perimeter

	Perimeter																
Number of squares		4	6	8	10	12	14	16	18	20	22	24	26	28	30	32	34	36
	1	✓																
	2		✓															
	3			✓														
	4			✓	✓													
	5				✓	✓												
	6				✓	✓	✓											
	7					✓	✓	✓										
	8					✓	✓	✓	✓									
	9					✓	✓	✓	✓	✓								
	10						✓	✓	✓	✓	✓							
	11						✓	✓	✓	✓	✓	✓						
	12						✓	✓	✓	✓	✓	✓	✓					
	13							✓	✓	✓	✓	✓	✓	✓				
	14							✓	✓	✓	✓	✓	✓	✓	✓			
	15							✓	✓	✓	✓	✓	✓	✓	✓	✓		
	16							✓	✓	✓	✓	✓	✓	✓	✓	✓	✓	
	17								✓	✓	✓	✓	✓	✓	✓	✓	✓	✓

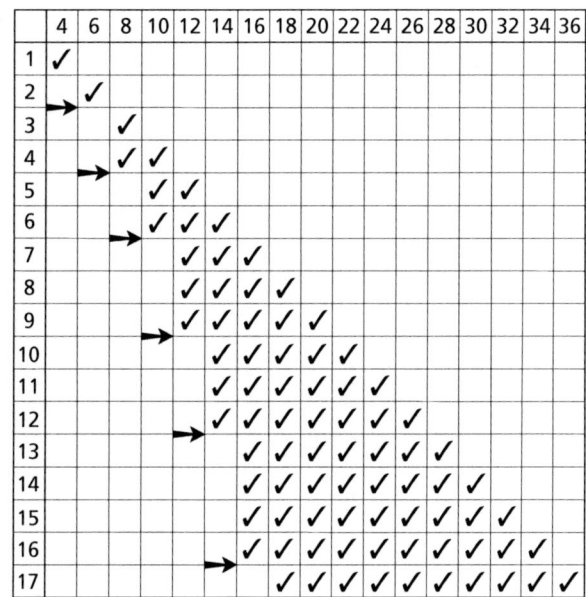

The minimum perimeter is shown by the left-hand edge of the block of ticks. This relationship is not linear but pupils may be able to give their own explanations of why the steps shown with arrows in the chart occur.

B **Rectangles** (p 196)

C **Square metres** (p 198)

> Optional: Newspapers

◊ Pupils often calculate answers in square metres with no idea how big a square metre is. At this stage, it is worthwhile making a metre square from newspaper and referring to it in working out or estimating the area of, say, a classroom wall.

It is also worth doing some work on estimating, say, the amount (and cost) of grass seed or turf needed to make an actual lawn (or sports field!). In 2003, a 1 kg box of grass seed sufficient for 20 square metres of grass cost about £6 to £8, and turf cost about £1.75 per square metre.

◊ From C3 onwards, working out missing lengths of sides is included. A sketch diagram is always a good idea and should be encouraged.

D **Bringing in triangles** (p 201)

E Using decimals (p 202)

◊ This section assumes knowledge that a half is 0.5 and a quarter is 0.25. It aims to help pupils see that the result of a decimal multiplication corresponds to the total of whole, half and quarter squares in an area diagram. You may need to check that pupils are following the ideas, through teacher-led class discussion.

Safely grazing (p 203)

This investigation may be done at any suitable time during the unit.

For pupils who have become confident with decimals, 'Safely grazing' can be adapted to start from a total fence length that gives rise to non-integer sides for the pen.

B Rectangles (p 196)

B1 12 cm²; the pupil's drawings of 1 cm by 12 cm, 2 cm by 6 cm rectangles

B2 The pupil's three rectangles such as
1 cm by 40 cm
2 cm by 20 cm
4 cm by 10 cm
5 cm by 8 cm

B3 1 cm by 16 cm
2 cm by 8 cm
4 cm by 4 cm (You can point out that a square is a special case of a rectangle.)

B4 (a) A 12 cm²   B 20 cm²
(b) A 16 cm   B 18 cm

B5 (a) 15 cm²   (b) 16 cm²   (c) 20 cm²
(d) 8 cm²   (e) 18 cm²

B6 (a) The pupil's answer (many people think more than half is shaded).
(b) The area of the whole rectangle is 42 cm²; the area of the shaded part is 20 cm². So less than half is shaded.

B7 (a) 7 cm   (b) 7 cm   (c) 4 cm
(d) 8 cm

B8 (a) 27 cm²   (b) 24 cm

C Square metres (p 198)

C1 (a) 8 m²   (b) 4 m²   (c) 12 m²
(d) 16 m

C2 (a), (b) The pupil's sketches leading to a total area of 12 m² each time

C3 (a) 17 m²   (b) 19 m²   (c) 15 m²
(d) 42 m²   (e) 18 m²

C4 (a) 18 m   (b) 22 m   (c) 18 m
(d) 28 m   (e) 22 m

C5 A 240 m², B 144 m², C 280 m²
D 240 m², E 400 m², F 24 m²
G 480 m², H 112 m², I 72 m²
In each case, the total area is 664 m².

C6 (a) 1275 m²   (b) 2256 m²

C7 A 352 m², B 48 m², C 63 m²
So the area of the green shape is 241 m².

C8 (a) 582 m²   (b) 480 m²

D Bringing in triangles (p 201)

D1 (a) A half   (b) 3 cm²

D2 (a) 10 cm²   (b) 7 cm²   (c) 1 cm²
(d) 1.5 cm²   (e) 7.5 cm²

D3 (a) $7.5\,\text{cm}^2$ (b) $13\,\text{cm}^2$ (c) $5\,\text{cm}^2$
(d) $6\,\text{cm}^2$ (e) $7\,\text{cm}^2$ (f) $4\,\text{cm}^2$
(g) $2\,\text{cm}^2$

D4 (a) $6\,\text{cm}^2$ (b) $8\,\text{cm}^2$ (c) $6\,\text{cm}^2$

E Using decimals (p 202)

E1 (a) $10.5\,\text{cm}^2$ (b) $9\,\text{cm}^2$ (c) $7\,\text{cm}^2$

E2 (a) $13\,\text{cm}$ (b) $13\,\text{cm}$ (c) $11\,\text{cm}$

E3 (a) $6\,\text{cm}^2$ (b) $12.5\,\text{cm}^2$ (c) $10\,\text{cm}^2$
(d) $4.5\,\text{cm}^2$ (e) $18\,\text{cm}^2$ (f) $22.5\,\text{cm}^2$

E4 (a) A quarter (b) 0.25 (c) $8.75\,\text{cm}^2$

E5 (a) $5.25\,\text{cm}^2$ (b) $11.25\,\text{cm}^2$
(c) $15.75\,\text{cm}^2$

E6 (a) $10\,\text{cm}$ (b) $14\,\text{cm}$ (c) $16\,\text{cm}$

E7 (a) $3.75\,\text{cm}^2$ (b) $13.75\,\text{cm}^2$
(c) $22.75\,\text{cm}^2$ (d) $0.75\,\text{cm}^2$
(e) $6.25\,\text{cm}^2$ (f) $8.25\,\text{cm}^2$

E8 (a) $9\,\text{cm}$ (b) $12\,\text{cm}$ (c) $4.5\,\text{cm}$
(d) $18\,\text{cm}$ (e) $36\,\text{cm}$

Safely grazing (p 203)

The biggest pen is 10 m by 10 m, with an area of $100\,\text{m}^2$.

What progress have you made? (p 204)

1 (a) The pupil's sketch giving area $16\,\text{m}^2$
(b) $20\,\text{m}$

2 (a) $7.5\,\text{cm}^2$ (b) $3\,\text{cm}^2$
(c) $2\,\text{cm}^2$ (d) $3.5\,\text{cm}^2$

3 (a) Area $3.75\,\text{cm}^2$, perimeter $8\,\text{cm}$
(b) Area $12.25\,\text{cm}^2$, perimeter $14\,\text{cm}$

Practice booklet

Sections A, B and C (p 61)

1 (a) 3 cm by 7 cm
The area is $21\,\text{cm}^2$;
the perimeter is 20 cm.
(b) 2 cm by 4 cm
The area is $8\,\text{cm}^2$;
the perimeter is 12 cm.

2 The pupil's sketches, giving areas
(a) $17\,\text{m}^2$ (b) $14\,\text{m}^2$ (c) $14\,\text{m}^2$
(d) $19\,\text{m}^2$ (e) $20\,\text{m}^2$ (f) $10\,\text{m}^2$

3 (a) $18\,\text{m}$ (b) $18\,\text{m}$ (c) $18\,\text{m}$
(d) $20\,\text{m}$ (e) $22\,\text{m}$ (f) $14\,\text{m}$

4 The pupil's sketches, giving areas
(a) $324\,\text{m}^2$ (b) $1400\,\text{m}^2$ (c) $706\,\text{m}^2$
(d) $231\,\text{m}^2$

Section D (p 63)

1 (a) $3\,\text{cm}^2$ (b) $4\,\text{cm}^2$ (c) $4.5\,\text{cm}^2$

2 The pupil's sketches, giving areas
(a) $210\,\text{m}^2$ (b) $72\,\text{m}^2$ (c) $252\,\text{m}^2$

Section E (p 64)

1 (a) $5\,\text{cm}^2$ (b) $5.25\,\text{cm}^2$ (c) $20.25\,\text{cm}^2$

2 (a) $9\,\text{cm}$ (b) $10\,\text{cm}$ (c) $18\,\text{cm}$

3 (a) $2.25\,\text{cm}^2$ (b) $16.5\,\text{cm}^2$ (c) $29.25\,\text{cm}^2$
(d) $9\,\text{cm}^2$ (e) $11.25\,\text{cm}^2$ (f) $11.25\,\text{cm}^2$

Review 3 (p 205)

Essential
Angle measurer, compasses, calculator

1. (a) $w = 2r + 2$ (b) 42
2. (a) A: 36 m B: 24 cm
   (b) A: 56 m² B: 24 cm²
3. $a = 62°$, $b = 30°$, $c = 30°$, $d = 150°$, $e = 60°$, $f = 60°$
4. The third, pink triangle
5. The pupil's drawing of a rectangle 4 cm by 12 cm
6. (a) $\frac{1}{2}$ (b) $\frac{1}{3}$ (c) $\frac{5}{6}$
   (d) $\frac{3}{4}$ (e) $\frac{7}{12}$
7. (a) The pupil's accurate drawings
   (b) AB = 9.5 cm, BC = 3.9 or 4 cm, angle at C = 78°
   PR = 8.6 or 8.7 cm, angle at Q = 60°, angle at R = 30°
   (c) PQR is scalene.
   ABC is isosceles; PQR is scalene and right-angled.
8. $\frac{2}{3}$
9. (a) 7.17 kg (b) 0.53 kg
   (c) 4 kg (d) 4.1 kg
   (e) Phoebe: 6.74 kg
   Hayley: 7.6 kg
   Jennifer: 7.04 kg
   Kate: 8.1 kg
10. 0.3 kg
11. (a) The pupil's accurate drawing of a rectangle 6.5 cm by 2.5 cm
    (b) $16\frac{1}{4}$ cm² or 16.25 cm²
    (c) 16 cm²
12. 0.33 kg
13. £2.40

Mixed questions 3 (Practice booklet p 65)

1. (a) 73 kg (b) 7 kg (c) 67 kg
   (d) 9 kg
2. (a)

Pocket money	Frequency
£1.00–£1.99	2
£2.00–£2.99	5
£3.00–£3.99	10
£4.00–£4.99	7
£5.00–£5.99	3

   (b) The pupil's grouped frequency chart for the data above
   (c) £3.00–£3.99 (d) 7
3. (a) 6.4 (b) 14.7 (c) 6.9 (d) 9.1
4. (a) 14.28 km (b) 2.88 km
5. $a = 19°$ $b = 161°$
6. £0.41 or 41p
7. $\frac{3}{5}$
8. The pupil's right-angled scalene triangle with an area of 8 cm²
9. (a) 35.78 (b) 2.29 (c) 0.67
10. (a) 202 (b) $w = 2g + 2$
    (c) The pupil's explanation: for example, 'Each grey tile has two adjacent white tiles and then there are two extra white tiles at the ends.'
11. 1.68 m
12. (a) $1\frac{2}{5}$ (b) $2\frac{1}{10}$ (c) $1\frac{2}{13}$ (d) $1\frac{7}{9}$
13. Area 24 m², perimeter 26 m

27 Spot the rule

7C/19

The teacher-led, whole-class activity that starts section A is the most important activity in this unit.

At first sight, it may seem perverse not to use input numbers in order, increasing by 1. But this encourages just one method for solving these rule puzzles. An equally effective method may be to ask 'What happens to 100? to 1000?', and pupils may use others.

p 207 **A** Finding rules	Finding a rule that connects pairs of numbers
p 208 **B** Using letters	Finding a rule from a table Using *n* to stand for *number*
p 210 **C** More shorthand	Using shorthand such as 4*n* and $\frac{n}{2}$

Practice booklet pages 67 to 69

A Finding rules (p 207)

◊ This section starts with a teacher-led activity with the whole class. This is especially engaging and effective if carried out in complete silence, as the following account from a school makes clear.

'It was the last lesson on Friday afternoon. I allowed time for settling down and said we were going to play a game ... everyone had to be silent, including me!

Without saying anything, I wrote on the board ...

    2 → 9
    5 → 12
    3 →

... and (still without saying anything) invited someone to come and fill in the gap.

    3 → 10 ☺

As they were right I drew a smiley face.

I then added more numbers each time on the left-hand side. Different pupils came and wrote up numbers on the right-hand side.

3	→	10	☺		
7	→	15	☹	14	☺
10	→	17	☺		
0	→	7	☺		
25	→				

If the pupil's number was wrong I drew a sad face and signalled another pupil to try. They liked big numbers, so I continued with larger ones.

After a while I wrote up …

number →

… and received the responses

number	→	add 7	☹
	→	number add on 7	☺

I was looking for the rule here (which could be in words).

After several numbers had been written on the board and pupils had come up and answered in turn, I wrote up a selection of numbers in the left-hand column which included some easy numbers and some more difficult ones.

Pupils came out and filled in the ones they felt comfortable with.

Decimals and negative numbers were included when appropriate.

2	→	9	
5	→	12	
25	→		
109	→		
4	→	11	☺
964	→		
0.5	→	7.5	☺
⁻3.4	→		

After a few games, instead of *number* I introduced the class to using *n* or another letter. I think it is important to emphasise the use of *n* for number when pupils are expressing their rules. Initially $n \rightarrow + 3$ is quite common instead of $n \rightarrow n + 3$.

Once I had played three or four games with the class I asked the pupils to think up some rules of their own. They then took turns to try out their games with the class.

I found there was a beautiful atmosphere with the lesson progressing with no one ever speaking. This can happen quite naturally and seems to strengthen pupils' focus on puzzling out the rule. It allows time for everyone to think and not just the quick ones.'

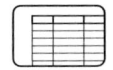

◊ After playing the game with the whole class, you could continue in groups.
A variation is to set up a simple spreadsheet to produce outputs automatically – nothing complicated, just a formula such as = A1*2 + 1 hidden in cell B1. Pupils can then just type numbers into A1 and see each corresponding result in B1. You could use graphic calculators in a similar way.

B Using letters (p 208)

C More shorthand (p 210)

A Finding rules (p 207)

A1 (a) 31   (b) 10   (c) 22   (d) 34
(e) 1   (f) 301

A2 The pupil's results for
number → (*number* × 2) – 1

A3 The pupil's rule and results

B Using letters (p 208)

B1 (a)

$n \to n - 4$
7 → 3
12 → **8**
20 → **16**
4 → **0**
13 → 9
24 → **20**

(b)

$n \to n \times 5$
3 → 15
8 → **40**
2 → **10**
10 → **50**
4 → **20**
6 → **30**

(c)

$n \to n \div 2$
10 → **5**
16 → **8**
22 → **11**
0 → **0**
8 → 4
12 → 6

B2 (a) $n \to (n \times 3) - 1$
(b) 29   (c) 11   (d) 2   (e) 32

B3 (a) $n \to (n - 2) \times 2$
(b) 6   (c) 8   (d) 20   (e) 0

B4 (a) It could be any of the rules.
(b) $n \to (n \times 3) - 4$

B5 (a) $n \to (n \times 4) - 8$
(b) The pupil's numbers following the rule

B6 The pupil's different rules for 3 → 6

B7 (a) *number* → half of *number*
$n \to n \div 2$

(b)

8 →	4
4 →	2
12 →	6
3 →	$1\frac{1}{2}$ or **1.5**
2 →	**1**
1 →	$\frac{1}{2}$ or **0.5**

B8 (a)

$n \to n \div 3$
12 → **4**
21 → **7**
30 → **10**
3 → **1**

(b)

$n \to n \times 4$
3 → **12**
5 → **20**
1 → **4**
10 → **40**

(c)

$n \to n \div 4$
8 → **2**
12 → **3**
20 → **5**
2 → $\frac{1}{2}$ or **0.5**

B9 (a) 3   (b) 6   (c) 1
(d) 10   (e) $2\frac{1}{2}$ or 2.5   (f) $\frac{1}{2}$ or 0.5

27 Spot the rule • 155

B10 (a) $n \to (n-2) \div 3$ or
number $\to$ (number $- 2) \div 3$
(b) 2 (c) 4 (d) 1
(e) 10 (f) 0

C More shorthand (p 210)

C1 (a)
$n \to n + 11$
3 → 14
10 → 21
30 → 41
0 → 11
56 → 67
32 → 43

(b)
$n \to n \div 5$
15 → 3
10 → 2
20 → 4
0 → 0
50 → 10

(c)
$n \to 2n + 1$
3 → 7
6 → 13
10 → 21
8 → 17
5 → 11

(d)
$n \to 10 - n$
3 → 7
8 → 2
4 → 6
9 → 1
5 → 5
7 → 3

C2 The pupil's rule using n

C3 (a) It could be any of the rules.
(b) $n \to 4n - 1$

C4 (a) $n \to 5n + 1$
(b) 21 (c) 101 (d) 126 (e) 1

C5
$n \to 4n - 2$
3 → 10
8 → 30
4 → 14
9 → 34
5 → 18

C6 (a) $n \to 3n + 2$ (b) $n \to 6n - 1$
(c) $n \to 2n + 3$

C7
$n \to \frac{n}{4}$
20 → 5
12 → 3
36 → 9
100 → 25
32 → 8

C8 (a) 2 (b) 5 (c) 2.5
(d) 0.5 (e) 0

C9 (a) $n \to 5n$ (b) $n \to 3n$
(c) $n \to 2n + 4$ (d) $n \to 6n + 2$
(e) $n \to 4n - 1$ (f) $n \to n + 12$
(g) $n \to \frac{n}{2}$ (h) $n \to 30 - 2n$
(i) $n \to \frac{n}{6}$ (j) $n \to 8 - 2n$
(k) $n \to 100 - 4n$ (l) $n \to \frac{n}{2}$

What progress have you made? (p 212)

1 (a)
$n \to n + 4$
3 → 7
6 → 10
10 → 14
4 → 8

(b)
$n \to n \times 3$
2 → 6
4 → 12
10 → 30
3 → 9

2 number $\to$ (number $\times$ 2) + 1 fits C.
number $\to$ number + 3 fits A.
number $\to$ (number $\times$ 3) $-$ 1 fits B.

3 number $\to$ (number $\times$ 2) $-$ 1
or $n \to 2n - 1$

4 (a) The pupil's four number pairs for $n \to n + 7$
(b) The pupil's four number pairs for $n \to n \times 10$

5 (a) (i) The pupil's four number pairs for $n \to 2n - 1$
(ii) The pupil's four number pairs for $n \to 3n + 10$
(b) (i) $n \to 2n + 1$
(ii) $n \to 4n - 10$

156 • 27 Spot the rule

Practice booklet

Sections A and B (p 67)

1. (a) 30 (b) 40 (c) 70
   (d) 110 (e) 140 (f) 0

2. (a) 7 (b) 11 (c) 21
   (d) 5 (e) 3 (f) 27

3. (a)

number → number + 5
n → n + 5
4 → **9**
2 → **7**
13 → **18**
3 → 8
20 → 25

   (b)

number → number ÷ 3
n → n ÷ 3
6 → **2**
15 → **5**
21 → **7**
30 → 10
27 → 9

4. (a) $n \to (n \times 2) + 5$
   (b) 9 (c) 13 (d) 25

5. (a) $n \to (n - 2) \times 3$
   (b) $n \to n \div 4$
   (c) $n \to n + 12$
   (d) $n \to (n \times 3) + 7$

6. (a) (i) $n \to n - 4$ (ii) 9, 15
   (b) (i) $n \to n \div 2$ (ii) 3, 14

7. (a) All five of them could be her rule.
   (b) $n \to (n \times 2) + 6$

8. (a) $n \to (n \times 3) - 3$
   (b) $10 \to 27$, $4 \to 9$, $1 \to 0$

Section C (p 69)

1. (a)

$n \to n - 7$
10 → **3**
20 → **13**
8 → **1**
17 → 10

   (b)

$n \to 3n$
4 → **12**
5 → **15**
13 → **39**
0 → 0

   (c)

$n \to 8 - n$
3 → **5**
6 → **2**
$\frac{1}{2} \to \mathbf{7\frac{1}{2}}$
2 → 6

2. $n \to 4n + 3$

3. (a)

$n \to 10n - 2$
3 → **28**
4 → **38**
7 → **68**
10 → 98

   (b)

$n \to \frac{n}{3}$
15 → **5**
21 → **7**
60 → **20**
27 → 9

4. (a) $n \to (n \times 2) + 1$ or $n \to 2n + 1$
   (b) $n \to (n \times 4) - 1$ or $n \to 4n - 1$
   (c) $n \to (n \times 5) + 1$ or $n \to 5n + 1$

5. (a) $n \to 7n$ or $n \to 7 \times n$
   (b) $n \to \frac{1}{2}n - 1$ or $n \to n \div 2 - 1$ or $n \to \frac{n}{2} - 1$
   (c) $n \to 5n - 3$ or $n \to (5 \times n) - 3$
   (d) $n \to \frac{n}{2} + 10$ or $n \to (2 \times n) + 10$

27 Spot the rule

28 Chance

7T/30, 7C/20

This unit introduces probability through games of chance. Probabilities are based on equally likely outcomes.

T	p 213 **A** Chance or skill?	Deciding whether a game involves chance or skill
T	p 214 **B** Fair or unfair?	Deciding whether a game of chance is fair
T	p 215 **C** Probability	The probability scale from 0 to 1
T	p 216 **D** Equally likely outcomes	Writing a probability as a fraction Finding the probability of an event not happening

p 218 **E** Choosing at random

Essential

Dice, counters
Sheets 111 to 115
Practice booklet pages 70 and 71

Optional

OHP transparencies of sheets 114 and 115

A Chance or skill? (p 213)

Dice and counters
Sheets 111 to 113 (game boards)

◊ Before discussing and playing the games, you could get pupils talking about chance, e.g. the National Lottery. People often have peculiar ideas about chance. For example, would they write on a National Lottery ticket the same combination as the one which won last week? If not, why not?

You could ask pupils to think about games they know and to discuss the elements of chance and skill in them.

◊ Before playing each game, ask pupils to try to decide from its rules whether it is a game of pure chance, a game of skill, or a mixture.

Some games of skill give an advantage to the first player. Who goes first is usually decided by a process of chance.

158 • 28 *Chance*

> 'For homework, I got them to devise a game themselves and say whether it was based on chance or skill. While some games produced were rather poor, this did produce some really excellent work from some pupils.'

◊ You could split the class into pairs or small groups, with each group playing one of the games and reporting on it.

◊ 'Fours' is a game of skill. 'Line of three' is a mixture of chance and skill. 'Jumping the line' appears to involve skill, because you have to decide which counters to move and it looks as if you can get 'nearer' to winning. But it is a game of pure chance. At any stage there is only one number which will enable the player to win. If any other number comes up, whatever the player does leaves the opponent in essentially the same position.

B Fair or unfair? (p 214)

> Dice, counters, sheets 114 and 115
> Optional: Transparencies of sheets 114 and 115

◊ You could start by playing 'Three way race' several times as a class, with a track on the board.

When pupils play the game themselves, ask them to record the results and then pool the class's results.

Let pupils consider each other's ways of making the game fairer (if they can think of any!). Do they agree that they would be fairer?

> 'The second race works well as a class if a rat number is assigned to a small group of pupils.'

> 'We had volunteer rats, a bookie, and the rest were punters as the rats moved across the room.'

◊ The first rat race is straightforward (although there may be some pupils who think that 6 is 'harder' to get than other numbers). In the second race you could ask for suggestions for making it fairer, still using two dice. (For example, the track could be shortened for the 'end' numbers; but even so, rat 1 is never going to win!)

For the second rat race, pupils could list possible outcomes to discover that there are more ways to make 7 than there are to make 3, for example. So some scores are more likely than others.

C Probability (p 215)

◊ Explain first the meanings of the two endpoints of the scale. Something with probability 0 is often described as 'impossible'. However, there are different ways of being impossible and some of them have nothing to do with probability (for example, it is impossible for a triangle to have four sides). So it is better to say 'never happens'. Something with probability 1 always happens, or is certain to happen.

◊ Go through the events listed in the pupil's book and discuss where they go on the scale. The coin example leads to the other especially important point on the scale, $\frac{1}{2}$. Associate this with 'equally likely to happen or not happen', with fairness, 'even chances', etc.

◊ Keep the approach informal. The important thing is to locate a point on the appropriate side of $\frac{1}{2}$, or close to one of the ends when appropriate (for example, in the case of the National Lottery!).

D Equally likely outcomes (p 216)

◊ A spinner is useful for probability. It shows fractions in a familiar way.

D4 If the pupil's answer for (d) is $\frac{1}{3}$, then they have ignored the inequality of the parts.

Odds

Although some teachers would like to outlaw 'odds', this language is used a lot in the real world. So it may be better to explain the connection, and the difference, between probability and odds.

Bookmakers' odds make an allowance for profit and are not linked to probability in the simple way shown on the pupil's page. It is only 'fair odds' that are so linked.

E Choosing at random (p 218)

◊ In some cultures, raffles and all forms of gambling are disapproved of. But if there are no objections you could simulate a raffle in class.

◊ There are some misconceptions which are worth bringing into the open. Some people think that a 'special' number, like 1 or 100, is less likely to win than an 'ordinary' number (because there are fewer 'special' than 'ordinary' numbers).

E4 Part (e) assumes knowledge of factors. You may wish to check pupils' knowledge before they try this question.

E6, 7 These questions should lead to discussion. Pupils may not have a strategy for comparing fractions, but may still give valid reasons for their choices. For example, 'B has twice as many reds as A but more than twice as many greens, so it's worse'.

'E6 and E7 were challenging for the top of a middle group but provoked interesting discussion.'

In E7, pupils may say 'Choose D, because it has 3 more greens and only 2 more reds'. The choice is correct but the reasoning is not. Suppose, for example, bag C had 2 green and 1 red and bag D had 5 green and 3 red. The probability of choosing green from C would be $\frac{2}{3}$, and from D $\frac{5}{8}$. $\frac{2}{3}$ is greater than $\frac{5}{8}$, so C would be the better choice.

C Probability (p 215)

C1

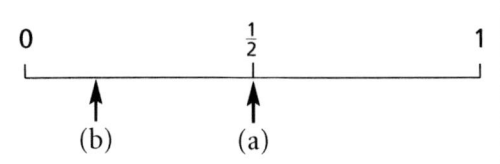

0 —————— 1/2 —————— 1
         ↑(b) ↑(a)

C2 (a) A boy
(b)

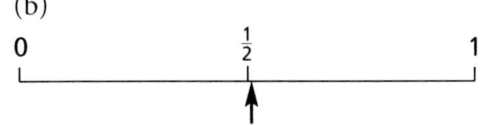

0 —————— 1/2 —————— 1
              ↑

D Equally likely outcomes (p 216)

D1 (a) $\frac{1}{2}$ (b) $\frac{1}{4}$ (c) $\frac{1}{6}$ (d) $\frac{1}{8}$

D2 (a) 1 (b) 0

D3 $\frac{2}{5}$

D4 (a) $\frac{2}{6}$ or $\frac{1}{3}$ (b) $\frac{3}{8}$ (c) $\frac{5}{8}$ (d) $\frac{1}{4}$

D5 (a) $\frac{1}{6}$ (b) $\frac{2}{6}$ or $\frac{1}{3}$ (c) $\frac{3}{6}$ or $\frac{1}{2}$

D6 (a) $\frac{5}{6}$ (b) $\frac{5}{8}$ (c) $\frac{3}{8}$ (d) $\frac{3}{4}$

D7 $\frac{3}{5}$

D8 (a) $\frac{2}{3}$ (b) $\frac{1}{8}$ (c) $\frac{4}{9}$ (d) $\frac{7}{10}$ (e) $\frac{1}{2}$

E Choosing at random (p 218)

E1 $\frac{1}{50}$

E2 $\frac{4}{200} = \frac{1}{50}$

E3 $\frac{1}{25}$

E4 (a) $\frac{1}{8}$ (b) $\frac{1}{4}$ (c) $\frac{3}{8}$ (d) $\frac{5}{8}$
(e) $\frac{1}{2}$ (f) 0 (g) $\frac{3}{4}$

E5 (a) $\frac{1}{100}$ (b) $\frac{1}{65}$ (c) $\frac{1}{64}$

*E6 Sarah should choose bag A.
$\frac{3}{8} = \frac{9}{24}$ $\frac{6}{18} = \frac{1}{3} = \frac{8}{24}$

*E7 Dilesh should choose bag D.
$\frac{4}{7} = \frac{48}{84}$ $\frac{7}{12} = \frac{49}{84}$

What progress have you made? (p 220)

1 (a)

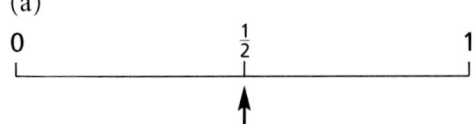

0 —————— 1/2 —————— 1
                  ↑

(b) It never happens.
(c) It always happens (or it is certain).
(d) See diagram.

2 (a) $\frac{2}{5}$ (b) $\frac{1}{4}$

3 $\frac{4}{80}$ or $\frac{1}{20}$

4 (a) $\frac{1}{6}$ (b) $\frac{2}{6}$ or $\frac{1}{3}$ (c) $\frac{4}{6}$ or $\frac{2}{3}$
(d) $\frac{1}{6}$ (e) $\frac{5}{6}$

Practice booklet

Sections D and E (p 70)

1 Spinner A
(a) $\frac{4}{8}$ or $\frac{1}{2}$ (b) $\frac{3}{8}$ (c) $\frac{1}{8}$
(d) $\frac{5}{8}$

Spinner B
(a) $\frac{2}{6}$ or $\frac{1}{3}$ (b) $\frac{1}{6}$ (c) $\frac{3}{6}$ or $\frac{1}{2}$
(d) $\frac{5}{6}$

Spinner C
(a) $\frac{1}{3}$ (b) $\frac{1}{3}$ (c) $\frac{1}{3}$
(d) $\frac{2}{3}$

2 (a) A (b) A (c) B

3 $\frac{1}{2}$

4 (a) $\frac{2}{9}$ (b) $\frac{1}{9}$ (c) $\frac{6}{9}$ or $\frac{2}{3}$
(d) $\frac{6}{9}$ or $\frac{2}{3}$ (e) $\frac{3}{9}$ or $\frac{1}{3}$ (f) $\frac{4}{9}$

*5 The probabilities for choosing a blue from each bag are

Bag P $\frac{3}{10}$ Bag Q $\frac{1}{3}$ Bag R $\frac{1}{4}$

So Rick should pick from bag Q as it has the highest probability of picking a blue.

28 Chance • 161

29 Oral questions: calendar (p 221) 7T/8

This page is for teacher-led oral work on interpreting a calendar.

◊ These sample questions are roughly in order of increasing difficulty.

1 What day of the week is 3 March in the year 2011? Thursday
2 Is 2011 a leap year? No
3 How many days are there in September? 30
4 What day of the week is Christmas day in 2011? Sunday
5 How many Saturdays are there in October in 2011? 5
6 Which months have five Thursdays in them?
   March, June, September, December
7 Whitsun bank holiday is the last Monday in May.
   What date is that? 30 May
8 The first day of a school trip is 22 June and the last day is 30 June.
   How many days is that? 9
9 How many shopping days (not Sundays) are there from 21 November until Christmas? 30
10 Karen sends a postcard on 9 August. It gets to her grandmother four days later. What date does it get there? 13 August
11 Ted gets a bill dated 5 October and has 21 days to pay.
   What is the latest date he can pay? 26 October
12 On 5 August, Cyril gets a parcel that was posted in Australia on 13 July. How many days did it take to arrive? 23
13 Peter gets some books out of the library on 27 May. They are due back three weeks later. What date are they due back? 17 June
14 The last day of Nicola's school term is 22 July. The first day back in the autumn is 5 September. How many weeks summer holiday does she get? 6
15 Ted hires some scaffolding on Monday 24 October.
   It is due back to the hire company on the Monday six weeks later.
   What date is it due back? 5 December
16 Councillor James holds a residents' advice session on the third Wednesday in every month. What dates does he hold it in July and August? 20 July, 17 August
17 Mrs Brown's home help comes on alternate Wednesdays, starting on the fifth of January. What dates does she come in March? 2, 16, 30 March

You can make work of this kind more immediate by working from an OHP photocopy of a current calendar and referring to actual dates of school events, sports fixtures and so on.

30 Number grids

7C/23

In this unit, pupils solve number grid problems using addition and subtraction. This includes using the idea of an inverse operation ('working backwards').

Algebra arises through investigating number grids. Pupils simplify expressions such as $n + 4 + n - 3$ and produce simple algebraic proofs of general statements.

p 222 **A** Square grids	Simple addition and subtraction problems
	Investigating 'diagonal rules' on number grids
p 224 **B** Grid puzzles	Using the idea of an inverse operation ('working backwards') to solve number puzzles
	Solving more difficult puzzles, possibly by trial and improvement
p 226 **C** Algebra on grids	Knowing that, for example, a number 2 more than n is $n + 2$
	Simplifying expressions such as $n + 4 - 3$
p 228 **D** Grid investigations	Exploring and describing number patterns
	Simple algebraic proofs
p 230 **E** Using algebra	Simplifying expressions such as $n + 4 + n - 3$
	Simple algebraic proofs

Optional
A4 sheets of paper
Felt-tip pens or crayons
Squared paper

Practice booklet pages 71 to 74

A Square grids (p 222)

The idea of a number grid is introduced. There are many opportunities to discuss mental methods of addition and subtraction.

Optional: Squared paper is useful for drawing grids; A4 sheets of paper and felt-tip pens or crayons (for 'Human number grids')

30 Number grids • 163

T

'Human number grids was a very helpful introduction – returned to it several times throughout unit.'

'This activity was an effective introduction to the idea of a number grid but was a bit of a nightmare last thing in the afternoon!'

Human number grids

This introductory activity does not appear in the pupil material.

◊ Each pupil or pair of pupils represents a position in a number grid.
Each position will contain a number. (For pupils familiar with spreadsheets, the idea of a 'cell' may help.)
The operations used are restricted to addition and subtraction.

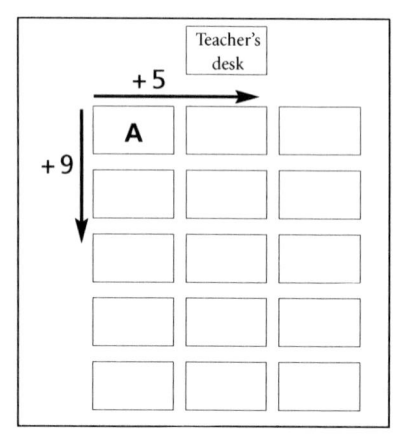

◊ Tables or desks need to be arranged in rows and columns so that the cells form a grid. Explain, with appropriate diagrams, that the class is going to form a human number grid that uses rules to get from a number in one cell to a number in another. A possible diagram is shown below.

◊ Referring to the numbers is easier if the cells are labelled.
Pupils can discuss how each cell might be labelled, for example:
  - A1, A2, B1, ... as on a spreadsheet
  - A, B, C, ...
  - or with the pupils' names

◊ Initially, it may be beneficial to use only addition or use sufficiently large numbers in cell A to avoid the complication of negative numbers.

◊ Decide on the first pair of rules and ask the pupils in the cell marked A in the diagram to choose a number for that cell.
Discuss how the numbers in other cells are found.
Now ask the pupils in cell A to choose another number, write it on both sides of a sheet of paper and hold it up.
Pupils now work out what number would be in their cell, write it on both sides of their sheet of paper and hold it up.
This can be repeated with different pupils deciding on the number for their cell.

◊ Questions can be posed in a class discussion, for example:
- Suppose the number in Julie and Asif's cell is 20.
  What number is in your cell, Peter?
  What number will be in Jenny's cell?
- What number do we need to put in cell A so that the number in cell F is 100?
- Find a number for cell A so that the number in cell K is negative.
- What happens if the 'across' and 'down' rules change places?

Ask pupils to explain how they worked out their answers. You could introduce the idea of an 'inverse' and encourage more confident pupils to use this word in their explanations.

Square grids

◊ Point out that all grids in the unit are square grids.

◊ One teacher presented unfinished grids on an OHP transparency and asked for volunteers to fill in any empty square. She found that less confident pupils chose easy squares to fill in while 'others with more confidence chose the hardest, leading to class discussion, and the idea of a "diagonal" rule came out naturally.'

◊ In one school, the class looked at rules in every possible direction as shown in the diagram.

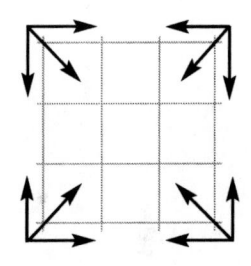

'Many were surprised by the fact that there was more than one route from one square to another, giving the same answer.'

◊ In discussion, bring out the fact that there are different ways to calculate a number in a square depending on your route through the grid. For example, the number in the bottom right-hand square in a 3 by 3 grid can be reached in six different ways. Pupils can try to find all these ways. Investigating the number of different routes through a grid to each square can lead to work on Pascal's triangle.

A1 In part (b), make sure pupils realise that the diagonal rule fits any position on these grids and not just those on the leading diagonal.
In part (c), some pupils may continue to use the '+ 6' and '+ 2' rules here. Discuss why using the diagonal rule '+ 8' could be used to give the same result.

A2 In part (b), emphasise that the diagonal rule fits any position on the grid.

A3 It is very likely that negative numbers will appear in the grids, providing an opportunity to consolidate work on negative numbers. However, if you want to try to avoid this, you could suggest that pupils choose quite large numbers for the top left-hand square of their grids or stick to rules that involve addition only.

30 Number grids

Appropriate teacher input is important here. Pupils could:
- consider rules that involve addition only
- choose two rules and investigate grids that use those rules only
- consider rectangular grids

Some pupils will easily see the link between the across, down and diagonal rules and can move quickly on to question A4.

A4 Some pupils may need to enter numbers in the grid to find the diagonal rules. Encourage pupils to use the results of their investigation in A3 to calculate the diagonal rules from the across and down rules only.

A5 Some pupils can look for all possible pairs, introducing the idea of an infinite number of pairs of the forms
- '+ a' and '+ $(11 - a)$' in part (a)
- '+ b' and '+ $(4 - b)$' in part (b)

B Grid puzzles (p 224)

◊ For B1 to B3, emphasise that it is not necessary to complete the whole grid to solve the puzzle, just find the missing number or rules.

◊ Encourage pupils to use the word 'inverse' to describe their methods.

B3 As a possible extension, pupils could make up their own puzzles like this to solve. However, puzzles without at least one pair of numbers in the same horizontal or vertical row (like (f)) are easy to construct but more difficult to solve. These may provide an enjoyable challenge, but if they lead to frustration, pupils could be restricted to making up puzzles with at least one pair of numbers in the same horizontal or vertical row.

Pupils could solve puzzle (f) using trial and improvement or possibly by the following more direct method.

The rule in this diagonal direction (↗) is '+ 5' so the number in the top row directly above the 6 is 10 + 5 = 15.

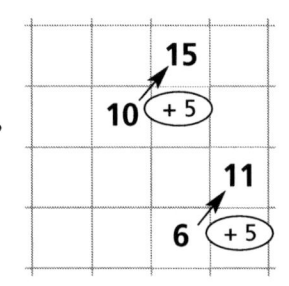

Now the down rule is easily found to be '– 3' and the across rule is '+ 2'.

This method can be adapted for the problems in B5 (adding more rows and columns to the grid where necessary).

*B5 Pupils can devise their own methods to solve these. Encourage them to be systematic and ask them to explain their methods to you or each other. Some may be able to solve these problems using a direct method such as the one above. Others may devise systematic trial and improvement methods.

*B6   Encourage pupils to be systematic in their choice of pairs to make the link easier to find. The link could be expressed in words: for example, 'If the across rule is to add a number, then the down rule is to subtract twice that number'.

Ⓒ Algebra on grids (p 226)

Algebra is introduced in the context of number grids.

◊ The teacher-led introduction begins with a grid that uses addition only. After discussion of this grid, you may wish pupils to try questions C1 to C3 where the rules are restricted to addition. Then move on to the second grid on page 226 and to questions C4 to C6. You may find a number line is helpful in getting these ideas across.

Remind pupils of earlier work in section A on finding rules. Emphasise that the expressions in the grid show how to find any number on the grid *directly* from the top left corner. For example, as the expression in the bottom right corner is $n + 14$, then the rule to go from the number in the top left square to the number in the bottom right square is '+ 14'.

In the second grid, pupils who suggest '$h - 9$' for the square below '$h - 2$' are possibly thinking of '$h - 2 + 7$' as '$h - (2 + 7)$'. Discussion of numerical examples may help to clear up any confusion.

Ensure that pupils understand, for example, that '$h + 7 - 4$' gives the same result as '$h - 4 + 7$'.

C2   Emphasise that the expression in the bottom right square gives a direct method of finding these numbers.

Ⓓ Grid investigations (p 228)

D1   In part (b) you may need to emphasise that the across and down rules must add and take away the *same* number.

◊ D3 to D6 provide an opportunity for more able pupils to choose to use algebra for themselves to explain their findings.

After pupils have investigated opposite corners for themselves, draw their conclusions together in a discussion that leads to the algebraic ideas in section E.

Ⓔ Using algebra (p 230)

◊ The teacher-led discussion of the use of algebra leads directly from the pupils' investigations in questions D3 to D6.

Ensure that pupils are aware that finding the expression for the opposite corners total to be $2n + 6$ each time proves that the totals will be the same for any value of n and also gives the rule to find this total.
Extend your discussion to consider how to simplify expressions such as $n - 3 + n - 4$ and $n + n + 2 + n - 3$.

E3 This gives an opportunity to emphasise the fact that equivalent expressions are equal for *all* values of the variables and not just a selected few.

As an extension, pupils could find and prove that the opposite corners total on a 3 by 3 grid is 2 times the centre number or that the diagonals total is 3 times the centre number.

A Square grids (p 222)

A1 (a) The pupil's grids

(b) The rule is '+ 8', with the pupil's explanations.

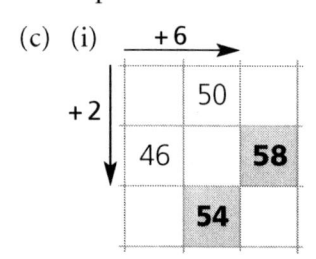

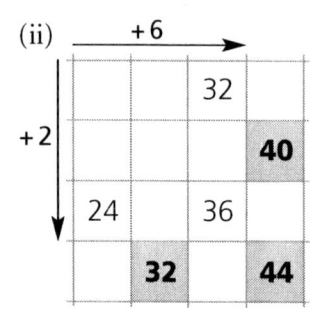

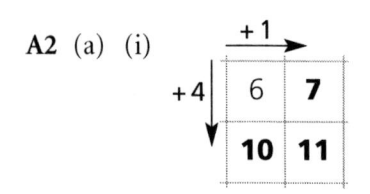

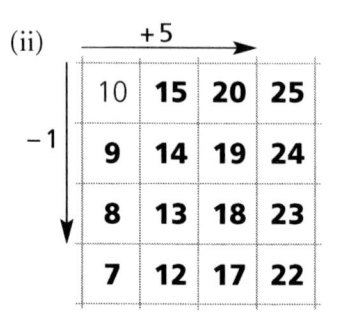

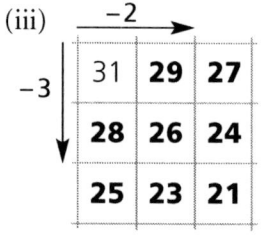

(c) (i)

+6 across, +2 down

	50	
46		**58**
	54	

(ii)

+6 across, +2 down

	32	
		40
24	36	
32		**44**

A2 (a) (i)

+1 across, +4 down

6	7
10	11

(ii)

+5 across, −1 down

10	15	20	25
9	14	19	24
8	13	18	23
7	12	17	22

(iii)

−2 across, −3 down

31	29	27
28	26	24
25	23	21

(b) (i) + 5   (ii) + 4   (iii) − 5
with the pupil's explanation

A3 The pupil's investigation and description of the link

A4 (a) + 17   (b) + 5

A5 (a) The pupil's pairs of rules equivalent to '+ 11', for example,

across '+ 1', down '+ 10'
or across '+ 5', down '+ 6'

(b) The pupil's pairs of rules equivalent to '+ 4', for example,

across '+ 1', down '+ 3'
or across '− 1', down '+ 5'

B Grid puzzles (p 224)

B1 (a) 23* (b) 30* (c) 39
(d) 25 (e) 43 (f) 95

*Pupils who have not grasped the idea that the rules operate from left to right and from top to bottom might give 27 and 90 as their answers for parts (a) and (b) respectively.

B2 (a) Across '+ 4', down '+ 9'
(b) Across '– 4', down '– 2'
(c) Down '+ 5' (d) Down '– 3'
(e) Down '+ 21' (f) Across '– 5'

B3 (a) Across '+ 3', down '+ 8'
(b) Across '+ 7', down '– 3'
(c) Across '– 3', down '– 4'
(d) Across '– 2', down '+ 11'
(e) Across '– 5', down '– 3'
(f) Across '+ 2', down '– 3'

B4 (a) (f) is usually the most difficult.
(b) The pupil's reasons

***B5** (a) Across '+ 1', down '+ 2'
(b) Across '+ 6', down '– 2'
(c) Across '– 1', down '+ 4'

***B6** (a) The pupil's pairs of rules, for example,
across '+ 1', down '– 2'
or across '– 4', down '+ 8'
(b) The pupil's descriptions of the link, for example,

'If the across rule is to add a number, then the down rule is to subtract twice that number.'

'If the across rule is to subtract a number, then the down rule is to add twice that number.'

C Algebra on grids (p 226)

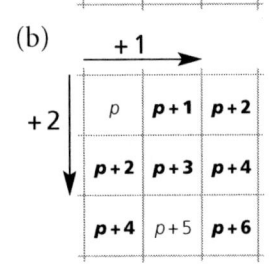

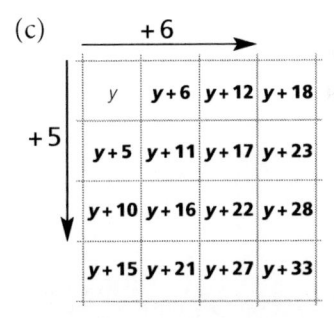

C1 (a)

	+3 →	
n	$n+3$	$n+6$
$n+4$	$n+7$	$n+10$
$n+8$	$n+11$	$n+14$

(+4 down)

(b)

	+1 →	
p	$p+1$	$p+2$
$p+2$	$p+3$	$p+4$
$p+4$	$p+5$	$p+6$

(+2 down)

(c)

	+6 →		
y	$y+6$	$y+12$	$y+18$
$y+5$	$y+11$	$y+17$	$y+23$
$y+10$	$y+16$	$y+22$	$y+28$
$y+15$	$y+21$	$y+27$	$y+33$

(+5 down)

C2 (a) 114 (b) 106 (c) 133

C3 (a) Across '+ 6', down '+ 5'
(b) Across '+ 5', down '+ 1'
(c) Across '+ 3', down '+ 10'

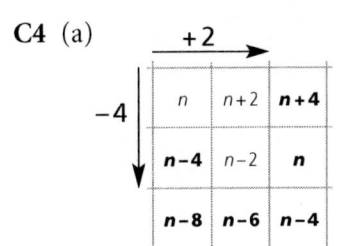

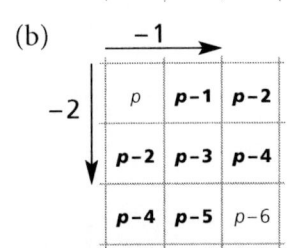

C4 (a)

	+2 →	
n	$n+2$	$n+4$
$n-4$	$n-2$	n
$n-8$	$n-6$	$n-4$

(–4 down)

(b)

	–1 →	
p	$p-1$	$p-2$
$p-2$	$p-3$	$p-4$
$p-4$	$p-5$	$p-6$

(–2 down)

(c)

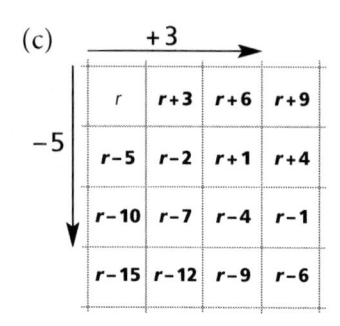

	+3 →			
↓ −5	r	r+3	r+6	r+9
	r−5	r−2	r+1	r+4
	r−10	r−7	r−4	r−1
	r−15	r−12	r−9	r−6

C5 (a) 54 (b) 56 (c) 56

C6 (a) $f + 11$ (b) $y + 12$ (c) $x + 6$
(d) $z - 10$ (e) $p + 1$ (f) $m - 3$
(g) $q + 3$ (h) $w - 14$ (i) $h + 5$

D Grid investigations (p 228)

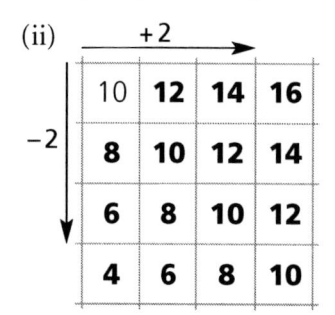

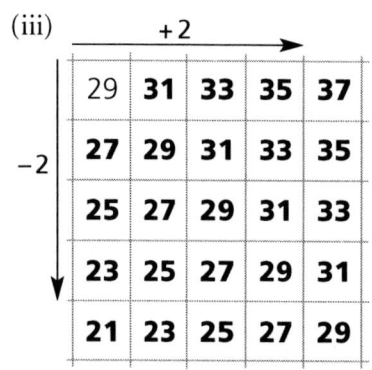

D1 (a) (i)

	+2 →		
↓ −2	24	26	**28**
	22	24	26
	20	22	24

(ii)

	+2 →			
↓ −2	10	12	14	16
	8	10	12	14
	6	8	10	12
	4	6	8	10

(iii)

	+2 →				
↓ −2	29	31	33	35	37
	27	29	31	33	35
	25	27	29	31	33
	23	25	27	29	31
	21	23	25	27	29

(b) The pupil's investigation
(c) The pupil's observations about symmetry and diagonals

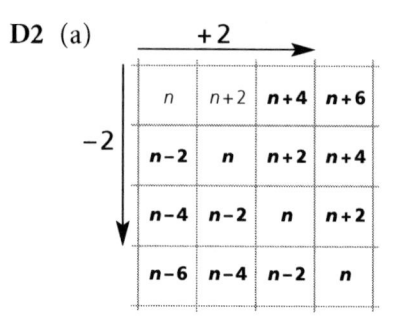

D2 (a)

	+2 →			
↓ −2	n	n+2	n+4	n+6
	n−2	n	n+2	n+4
	n−4	n−2	n	n+2
	n−6	n−4	n−2	n

(b) The pupil's explanation, for example, the expression in both those squares is the same (n) so the numbers will be the same.

(c) The pupil's observations, for example, the numbers in any diagonal going down from left to right are the same.

D3 (a) $37 + 13 = 50$ (b) The pupil's grids
(c) For each grid, the opposite corners totals are the same – this result could be explained using algebra.

D4 The pupil's investigation

D5 (a)

Opposite corners table	
Top left number	Opposite corners total
2	10
3	12
4	14
10	26

(b) The pupil's grids and results
(c) 'The opposite corners total is the top left number times 2 and add 6' or
'… (the top left number × 2) + 6' or
'… (the top left number + 3) × 2' or
'… $2n + 6$' or equivalent.
(d) 206

D6 The pupil's investigation

170 • 30 Number grids

E Using algebra (p 230)

E1 Grid P

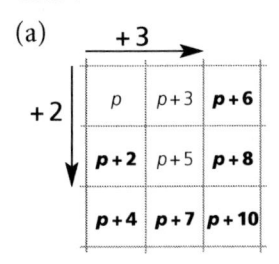

(a)

	+3 →		
+2 ↓	p	$p+3$	$p+6$
	$p+2$	$p+5$	$p+8$
	$p+4$	$p+7$	$p+10$

(b) Both pairs of corners add up to $2p + 10$.

(c) Yes

(d) 210

Grid N

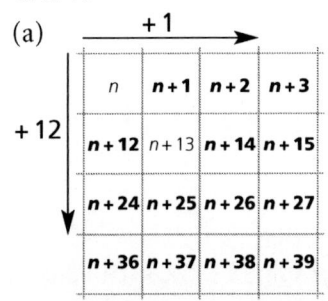

(a)

	+1 →			
+12 ↓	n	$n+1$	$n+2$	$n+3$
	$n+12$	$n+13$	$n+14$	$n+15$
	$n+24$	$n+25$	$n+26$	$n+27$
	$n+36$	$n+37$	$n+38$	$n+39$

(b) Both pairs of corners add up to $2n + 39$.

(c) Yes

(d) 239

Grid T

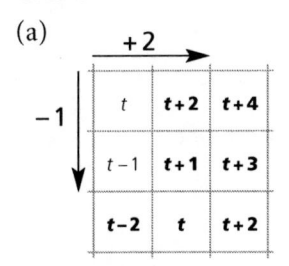

(a)

	+2 →		
−1 ↓	t	$t+2$	$t+4$
	$t-1$	$t+1$	$t+3$
	$t-2$	t	$t+2$

(b) Both pairs of corners add up to $2t + 2$.

(c) Yes

(d) 202

E2 A and I, B and G, C and E, D and H

E3 The pupil's explanation, for example, the expressions have the same value for **one** value of n but this does not mean that the expressions have the same value for **all** values for n.

E4 (a) $2p + 6$ (b) $2y + 9$
(c) $3q + 8$ (d) $3t + 4$
(e) $2x + 1$ (f) $3r + 6$
(g) $2w - 9$ (h) $2j - 1$
(i) $3h - 2$

E5 The pupil's investigation

What progress have you made? (p 232)

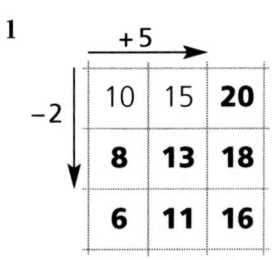

1

	+5 →		
−2 ↓	10	15	**20**
	8	**13**	**18**
	6	**11**	**16**

2 5

3 (a) Across '+ 5'
(b) Across '+ 6', down '− 2'
(c) Across '− 5', down '− 3'
(d) Across '− 4', down '+ 1'

4 (a) $n + 7$ (b) $p + 6$ (c) $y - 7$
(d) $t - 3$ (e) $2h + 3$ (f) $3v - 11$

Practice booklet

Sections A and B (p 71)

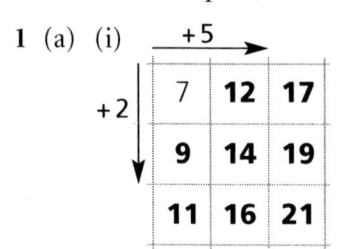

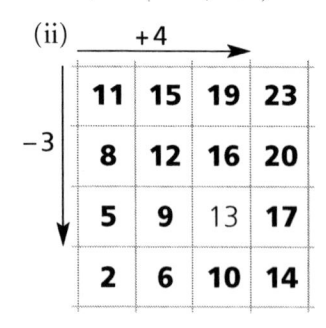

1 (a) (i)

	+5 →		
+2 ↓	7	12	17
	9	14	19
	11	16	21

(ii)

	+4 →			
−3 ↓	11	15	19	23
	8	12	16	20
	5	9	13	17
	2	6	10	14

(b) (i) + 7 (ii) + 1

2 (a) + 13 (b) + 2

Section C (p 72)

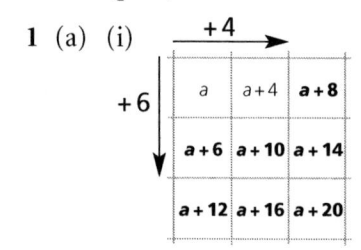

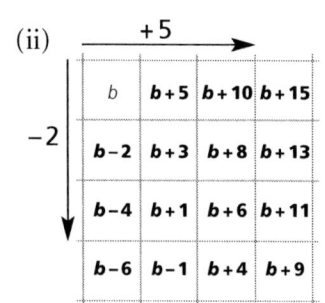

1 (a) (i)

	+4 →		
+6 ↓	a	a+4	a+8
	a+6	a+10	a+14
	a+12	a+16	a+20

(ii)

	+5 →			
−2 ↓	b	b+5	b+10	b+15
	b−2	b+3	b+8	b+13
	b−4	b+1	b+6	b+11
	b−6	b−1	b+4	b+9

(b) (i) 120   (ii) 109
(c) (i) 80   (ii) 91

2 (a) Across '+ 3', down '+ 4'
  (b) Across '+ 5', down '+ 3'
  (c) Across '− 1', down '+ 4'

3 (a) $t + 11$ (b) $a + 7$ (c) $q + 6$
  (d) $p + 4$ (e) $x + 10$ (f) $y + 2$
  (g) $s - 4$ (h) $v - 5$ (i) $b + 6$
  (j) $a - 4$ (k) $f - 16$ (l) $c + 2$
  (m) $d - 11$ (n) $g - 2$ (o) $h - 14$

Sections D and E (p 73)

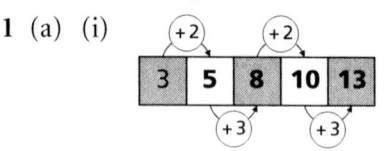

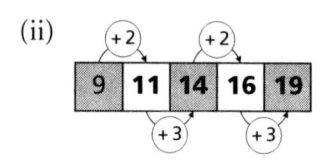

1 (a) (i) Chain: 3, 5, 8, 10, 13 with +2, +3, +2, +3

(ii) Chain: 9, 11, 14, 16, 19 with +2, +3, +2, +3

(b) The pupil's three completed chains

(c) The pupil's completed table with the first three numbers in the second column as 17, 13, 19

(d) The pupil's description of the rule such as 'To find the end number add 10 to the first number'. This may be expressed algebraically as, for example, $e = f + 10$.

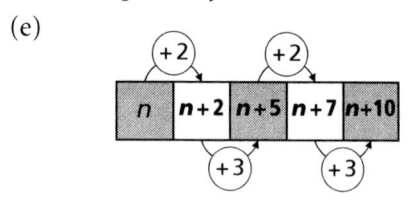

(e) Chain: n, $n+2$, $n+5$, $n+7$, $n+10$ with +2, +3, +2, +3

The last expression tells you that the end number is 10 more than the first number. It shows that the rule found in part (d) will work for any first number.

2 (a)

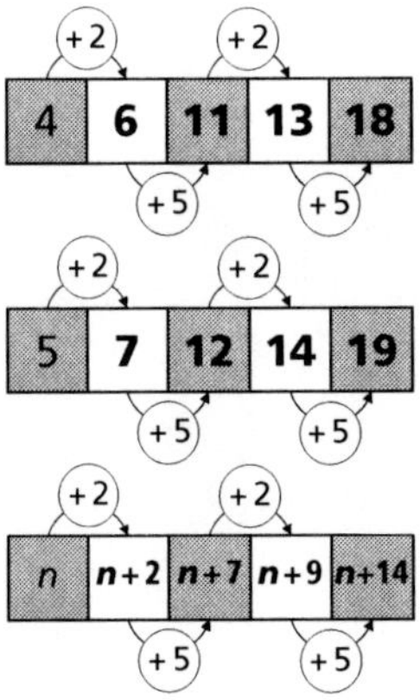
+2
+2
4
6
11
13
18
+5
+5
+2
+2
5
7
12
14
19
+5
+5
+2
+2
n
n+2
n+7
n+9
n+14
+5
+5

(b) 47

(c)

<u>Total table</u>

First number	Total
3	47
4	52
5	57
n	$5n + 32$

(d) The pupil's description of the rule such as 'To find the total multiply the first number by 5 and add 32'. This may be expressed algebraically as, for example, $t = 5n + 32$.

(e) 532

3 A and D, C and I, E and G

4 (a) $2p + 3$ (b) $3y + 9$ (c) $2q + 10$
(d) $2t + 3$ (e) $2x + 2$ (f) $3r + 6$
(g) $2w - 13$ (h) $2j - 3$ (i) $2h - 6$

31 Fractions and decimals

Work on a limited range of decimals and fractions between 0 and 1 provides a visual approach to equivalence and comparing sizes.

As presented here, the work for the first four sections is largely based on a resource sheet. An alternative approach, which some teachers have found very successful, is outlined at the end of the notes for section A.

A decimal is of course a fraction, but in this unit the common use of 'decimal' to mean decimal fraction and 'fraction' to mean vulgar fraction is followed.

T p 233 **A** Putting fractions on a fraction ruler

p 233 **B** Comparing fractions

T p 235 **C** Putting decimals on the fraction ruler

p 235 **D** Decimals and fractions

p 238 **E** Using mixed numbers

Essential

Sheet 150 (preferably enlarged on to A3 paper)
OHP transparency of sheet 150

Practice booklet pages 75 to 76

Optional

Centimetre squared paper, set squares and coloured pencils (for the alternative presentation)
Sheet 151 (to make decimal fans)

Ⓐ Putting fractions on a fraction ruler (p 233)

> Sheet 150, OHP transparency of sheet 150

T

◊ You could start by drawing a number line on the board and discussing where to label various points. Ask if pupils can label any points between 0 and 1, and explain that they will be zooming-in on the space between 0 and 1.

◊ On sheet 150, the space between 0 and 1 is represented by several scales, each divided into 100 equal parts. Check that pupils understand each division on this scale represents $\frac{1}{100}$. Ask pupils to label $\frac{10}{100}, \frac{20}{100}, \frac{30}{100}, \ldots$ under the scale marked 'hundredths'. Pick out some other points such as $\frac{46}{100}, \frac{4}{100}$ and $\frac{35}{100}$, and ask pupils to mark them above the scale with arrows and label them.

174 • 31 Fractions and decimals

Draw pupils' attention to the halves scale. How many halves can they label? Ask them to label $\frac{1}{2}$ below the scale. What is that in hundredths? Repeat this with the quarters, fifths, twentieths and tenths scales.

◊ As the labelling proceeds you could start to ask pupils to compare the sizes of fractions on different scales and to say which are equivalent.

◊ You could ask pupils to suggest other fractions to mark on the blank scales at the bottom (twenty-fifths and fiftieths are the obvious ones).

Alternative presentation

This approach is more time-consuming but teachers who have entered into the spirit of it have felt that pupils gained a greater sense of familiarity with the fractions.

> One sheet of centimetre squared paper, about 30 cm by 120 cm, for every two pupils
> Optional: Set square, coloured pencils

◊ Working in pairs, pupils first draw a line 100 cm long on the squared paper and label its ends 0 and 1. The larger format allows all of the hundredths to be labelled. Pupils can do this (perhaps working from opposite ends) in pencil first, so mistakes can be corrected.

Discuss the $\frac{50}{100}$ point and other ways it could be labelled. Decimals and percentages may arise as well as fractions. Now encourage pupils to think about other fractions they can mark. They should, for example, write $\frac{1}{4}$ below $\frac{25}{100}$, and put each 'family' of fractions (quarters, fifths and so on) on the same level. Colouring can help distinguish the families.

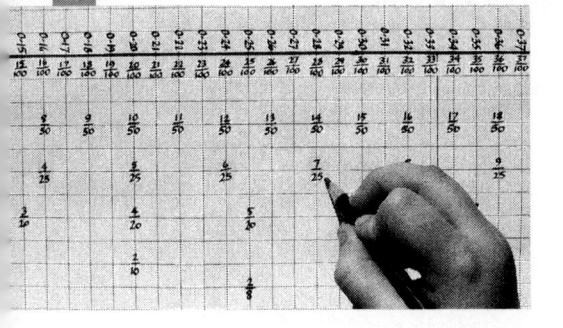

◊ Comparing the sizes of fractions and discussion of equivalence can then continue as with the resource sheet.

◊ Decimals can be added to the home-made fraction ruler as suggested in the guidance for section C.

B **Comparing fractions** (p 233)

> Sheet 150

◊ The fraction ruler is meant to help pupils make sense of equivalence and comparisons but they should eventually become independent of it. So if you start this section with oral questioning, the emphasis should be on answering a question with the ruler visible and then trying to answer a similar one with it out of sight.

C **Putting decimals on the fraction ruler** (p 235)

> Sheet 150, OHP transparency of sheet 150

◊ Remind pupils that another way of writing tenths is to use decimals. Ask them to label 0.1, 0.2, ... below the top scale. You can ask questions such as these (preferably with the fraction ruler now hidden) to check the basic idea is understood:
- What decimal is halfway between 0 and 1?
- How far is it between 0.2 and 0.6 as a decimal (and as a fraction)?

◊ Explain that decimals can also be used to write hundredths; for example, $\frac{46}{100}$ is 0.46. Ask pupils to mark these with arrows above the decimals scale and label them: 0.14, 0.36, 0.45, 0.78, … When they are confident about this ask them to label 0.10, 0.20, 0.30, … above the decimals scale.

◊ Referring to the fraction ruler, draw pupils' attention to the equivalences $0.50 = \frac{50}{100} = \frac{5}{10} = 0.5$. They can write their own sets of equivalences for 0.40 and so on.

◊ Ask how they should write $\frac{3}{100}$ as a decimal. If anyone says 0.3 (a common error), rather than saying this is wrong you could ask them what $\frac{30}{100}$ is as a decimal. Reflecting back a mistaken answer like this so that pupils have to sort out the resulting conflict often works better than simply telling them what is right and wrong. Make sure 0.03 is now correctly marked above the decimals scale.

Similarly, point to the mark halfway between 0 and 0.1 on the decimal scale and ask pupils to write it down as a decimal. If any write 0.5 ask them whether 0.5 is marked anywhere else on the sheet.

◊ To build confidence with two places of decimals, ask questions like
- What number is halfway between 0.8 and 0.9?
- How far is it from 0.34 to 0.4?

The emphasis should be on using the sheet for support then trying to answer similar questions without the sheet in sight.

◊ Also ask questions like
- Which is bigger, 0.39 or 0.7? (common error 0.39)
- Which is bigger, 0.09 or 0.2? (common error 0.09)

Here too, try a reflecting-back approach when errors occur.

D Decimals and fractions (p 235)

> Optional: Decimal fans made from sheet 151

Again, the aim is to make pupils independent of the sheet once it has served to build their confidence and understanding.

◊ The decimal fan (made by cutting out the pieces on sheet 151, piercing where marked and holding together with a brass fastener) is designed to promote instant recall of decimal equivalents. Pupils can each have a fan and you can ask (for example) 'Make a decimal equal to $\frac{4}{5}$'; pupils all show what they have made and you can soon see if anyone is having difficulty. Or you can make different decimals on two fans, hold them up to the class and ask which is bigger.

Less confident pupils have found it easier to use a fan made from just the left-hand set of cards on the resource sheet (0 to 9 and the decimal point).

E Using mixed numbers (p 238)

B Comparing fractions (p 233)

Answers given here are those available from the fraction ruler. Other correct answers may be possible.

B1 $\frac{10}{100}, \frac{2}{20}$

B2 (a) $\frac{25}{100}$ or $\frac{5}{20}$ (b) $\frac{6}{10}, \frac{60}{100}$ or $\frac{12}{20}$
    (c) $\frac{75}{100}$ or $\frac{3}{4}$ (d) $\frac{13}{20}$

B3 $\frac{4}{5}$

B4 $\frac{1}{20}, \frac{1}{2}, \frac{3}{5}, \frac{7}{10}$

B5 4

B6 (a) $\frac{2}{5} = \frac{8}{20}$ (b) $\frac{3}{5} = \frac{12}{20}$
    (c) $\frac{4}{5} = \frac{16}{20}$ (d) $\frac{1}{5} = \frac{4}{20}$

B7 5

B8 (a) $\frac{7}{20} = \frac{35}{100}$ (b) $\frac{3}{20} = \frac{15}{100}$
    (c) $\frac{4}{20} = \frac{20}{100}$ (d) $\frac{11}{20} = \frac{55}{100}$

B9 (a) $\frac{1}{2} = \frac{10}{20}$ (b) $\frac{3}{5} = \frac{6}{10}$
    (c) $\frac{4}{10} = \frac{8}{20}$ (d) $\frac{7}{10} = \frac{70}{100}$
    (e) $\frac{3}{4} = \frac{15}{20}$ (f) $\frac{30}{100} = \frac{6}{20}$
    (g) $\frac{18}{20} = \frac{90}{100}$ (h) $\frac{3}{5} = \frac{60}{100}$

B10 (a) $\frac{55}{100}$ (b) $\frac{3}{20}$ (c) $\frac{19}{50}$ (d) $\frac{1}{25}$

B11 $\frac{2}{5} = \frac{8}{20}, \frac{1}{4} = \frac{5}{20}$ so $\frac{2}{5}$ is bigger.

B12 Three fractions from $\frac{5}{10}, \frac{50}{100}, \frac{2}{4}, \frac{10}{20}$

B13 (a) $\frac{75}{100}$ or $\frac{15}{20}$ (b) $\frac{4}{10}$ or $\frac{40}{100}$ or $\frac{8}{20}$
    (c) $\frac{25}{100}$ or $\frac{1}{4}$ (d) $\frac{3}{10}$ or $\frac{6}{20}$

B14 $\frac{1}{2}$

B15 $\frac{20}{100}, \frac{3}{4}, \frac{9}{10}$

B16 2

B17 (a) $\frac{2}{10} = \frac{4}{20}$ (b) $\frac{6}{10} = \frac{12}{20}$
    (c) $\frac{7}{10} = \frac{14}{20}$ (d) $\frac{9}{10} = \frac{18}{20}$

B18 (a) $\frac{1}{2} = \frac{2}{4}$ (b) $\frac{1}{5} = \frac{2}{10}$ (c) $\frac{2}{5} = \frac{8}{20}$
    (d) $\frac{2}{10} = \frac{20}{100}$ (e) $\frac{1}{2} = \frac{5}{10}$ (f) $\frac{1}{4} = \frac{5}{20}$
    (g) $\frac{2}{5} = \frac{4}{10}$ (h) $\frac{15}{100} = \frac{3}{20}$

31 Fractions and decimals • 177

D Decimals and fractions (p 235)

Answers given here are those available from the fraction ruler. Other correct answers may be possible.

D1 The pupil's labelled arrows on the decimals scale

D2 Marked on the decimals scale:
(a) 0.85 (b) 0.15 (c) 0.75

D3 (a) 0.35 (b) 0.25 (c) 0.55

D4 Two fractions from each of these:
(a) $\frac{8}{10}, \frac{80}{100}, \frac{4}{5}, \frac{16}{20}$ (b) $\frac{55}{100}, \frac{11}{20}$
(c) $\frac{15}{100}, \frac{3}{20}$ (d) $\frac{9}{10}, \frac{90}{100}, \frac{18}{20}$

D5 (a) 0.6 (b) 0.9 (c) 0.2 (d) 0.7

D6 2

D7 (a) 0.6 (b) 0.2 (c) 0.4 (d) 0.8

D8 (a) 0.8 (b) 0.35 (c) 0.48 (d) 0.09

D9 0.05, 0.45, 0.5, 0.54, 0.6

D10 (a) $\frac{35}{100}$, 0.35 (b) $\frac{5}{100}$, 0.05
(c) $\frac{45}{100}$, 0.45 (d) $\frac{55}{100}$, 0.55

D11 (a) 0.1 (b) 0.95 (c) 0.9 (d) 0.3
(e) 0.92 (f) 0.2 (g) 0.75 (h) 0.07

D12 (a) $\frac{47}{100}$ (b) $\frac{99}{100}$ (c) $\frac{2}{100}$ (d) $\frac{77}{100}$

D13 0.04, 0.12, $\frac{1}{2}$, 0.7

D14 0.09, $\frac{62}{100}$, 0.71, $\frac{3}{4}$, 0.8

D15 (a) 0.77 (b) 0.79 (c) $\frac{4}{5}$

D16 He is not correct. 0.28 is $\frac{28}{100}$, 0.7 is $\frac{7}{10}$. So you cannot just compare the 28 and the 7 (or a similar explanation).

D17 2

D18 (a) $\frac{6}{100}$, 0.06 (b) $\frac{98}{100}$, 0.98
(c) $\frac{34}{100}$, 0.34 (d) $\frac{66}{100}$, 0.66

D19 4

D20 (a) $\frac{44}{100}$, 0.44 (b) $\frac{8}{100}$, 0.08
(c) $\frac{96}{100}$, 0.96 (d) $\frac{76}{100}$, 0.76

D21 (a) 0.65 (b) 0.45 (c) 0.95

D22 (a) 0.7 (b) 0.65 (c) 0.6 (d) 0.25

D23 (a) $\frac{92}{100}$ (b) $\frac{43}{100}$ (c) $\frac{29}{100}$ (d) $\frac{61}{100}$

D24 Two fractions from each of these:
(a) $\frac{6}{10}, \frac{60}{100}, \frac{3}{5}, \frac{12}{20}$ (b) $\frac{5}{10}, \frac{50}{100}, \frac{1}{2}, \frac{2}{4}, \frac{10}{20}$
(c) $\frac{75}{100}, \frac{3}{4}, \frac{15}{20}$ (d) $\frac{45}{100}, \frac{9}{20}$

D25 0.08, 0.34, 0.6, 0.91

D26 (a) Fifths (b) $\frac{3}{5}$
(c) Twentieths (d) $\frac{13}{20}$
(e) Quarters (f) $\frac{3}{4}$

E Using mixed numbers (p 238)

E1 (a) 2.75 (b) 1.6 (c) 5.9 (d) 10.5

E2 (a) $8\frac{1}{4}$ (b) $7\frac{1}{5}$ (c) $9\frac{2}{5}$ (d) $3\frac{17}{20}$
(Other equivalent fractions are possible.)

E3 $2\frac{1}{5}, 2\frac{1}{4}, 2\frac{13}{20}, 2\frac{3}{4}, 2\frac{9}{10}$

What progress have you made? (p 238)

1 For each part, one of the fractions given, or another equivalent fraction:
(a) $\frac{2}{10}, \frac{20}{100}, \frac{1}{4}$ (b) $\frac{3}{10}, \frac{30}{100}$
(c) $\frac{9}{20}$ (d) $\frac{80}{100}, \frac{4}{5}, \frac{16}{20}$

2 $\frac{3}{4}$

3 $\frac{1}{2}, \frac{3}{5}, \frac{7}{10}, \frac{15}{20}$

4 (a) $\frac{4}{10}$ (b) $\frac{7}{20}$ (c) $\frac{71}{100}$ (d) $\frac{3}{100}$
(or other equivalent fractions)

5 (a) 0.4 (b) 0.15
(c) 0.09 (d) 0.14

6 (a) 3.4 (b) 10.25
(c) 9.9 (d) 1.05

7 (a) $4\frac{1}{5}$ (b) $9\frac{3}{4}$ (c) $3\frac{1}{20}$ (d) $4\frac{11}{20}$
(or other equivalent fractions)

Practice booklet

Section B (p 75)

1 $\frac{30}{100}, \frac{6}{20}$ and $\frac{3}{10}$

2 (a) Any fraction equivalent to $\frac{7}{10}$ such as $\frac{14}{20}$ or $\frac{70}{100}$

  (b) Any fraction equivalent to $\frac{15}{20}$ such as $\frac{3}{4}$ or $\frac{75}{100}$

  (c) Any fraction equivalent to $\frac{60}{100}$ such as $\frac{6}{10}$ or $\frac{3}{5}$

  (d) Any fraction equivalent to $\frac{7}{20}$ such as $\frac{14}{40}$ or $\frac{21}{60}$

3 5

4 (a) $\frac{\mathbf{4}}{5} = \frac{20}{25}$ (b) $\frac{\mathbf{15}}{25} = \frac{3}{5}$

  (c) $\frac{1}{5} = \frac{\mathbf{5}}{25}$ (d) $\frac{10}{25} = \frac{\mathbf{2}}{5}$

5 (a) $\frac{2}{10}, \frac{2}{5}, \frac{2}{4}$ (b) $\frac{5}{10}, \frac{3}{4}, \frac{4}{5}$

  (c) $\frac{3}{5}, \frac{73}{100}, \frac{19}{20}$

6 (a) $\frac{17}{20} = \frac{\mathbf{85}}{100}$ (b) $\frac{3}{4} = \frac{15}{20}$ (c) $\frac{4}{5} = \frac{16}{20}$

  (d) $\frac{1}{5} = \frac{20}{\mathbf{100}}$ (e) $\frac{42}{50} = \frac{21}{\mathbf{25}}$ (f) $\frac{16}{25} = \frac{\mathbf{64}}{100}$

  (g) $\frac{7}{10} = \frac{35}{\mathbf{50}}$ (h) $\frac{1}{2} = \frac{25}{\mathbf{50}}$

7 (a) $\frac{3}{4}$ (b) $\frac{9}{20}$ (c) $\frac{15}{20}$ (d) $\frac{4}{10}$

Sections D and E (p 75)

1 Two fractions from each of these (or other equivalent fractions)

  (a) $\frac{3}{5}, \frac{6}{10}, \frac{12}{20}, \frac{60}{100}$

  (b) $\frac{9}{20}, \frac{45}{100}$

  (c) $\frac{19}{20}, \frac{95}{100}$

  (d) $\frac{1}{5}, \frac{2}{10}, \frac{4}{20}, \frac{20}{100}$

2 (a) 0.5 (b) 0.7 (c) 0.6 (d) 0.25

3 (a) 0.38 (b) 0.03 (c) 0.85 (d) 0.16

4 (a) 0.14, 0.22, $\frac{1}{4}$, 0.3

  (b) $\frac{12}{25}$, 0.56, $\frac{73}{100}$, 0.8

  (c) 0.08, $\frac{34}{100}$, 0.49, 0.6, $\frac{3}{4}$

  (d) 0.04, 0.28, $\frac{30}{100}$, 0.32, $\frac{9}{20}$

5 (a) 0.43   (b) 0.41   (c) 0.55

6 (a) 2.9 (b) 1.5 (c) 6.4 (d) 11.25

7 (a) $8\frac{2}{5}$ (b) $4\frac{1}{4}$ (c) $1\frac{13}{20}$ (d) $7\frac{3}{10}$
  or other equivalent fractions

8 $2\frac{4}{5}, 2\frac{9}{10}, 3\frac{3}{20}, 3\frac{19}{100}, 3\frac{1}{4}$

32 Oral questions: measures (p 239)   7T/28

The information table in the pupil's book is the basis for oral questions.

◊ Aim for regular sessions of oral questions. You can use the information table on several different occasions. Do not give too many questions: it is better to stop with all pupils feeling some sense of achievement after, say, five minutes rather than persist for half an hour!

◊ Start by explaining what is meant by the diameter of a ball, and that the diameter may vary slightly, which is why a range of values is given (for example, the men's shot can be any measurement between 11 and 13 cm in diameter). Similarly a range is given for the weight and cost in some cases.

◊ Some possible questions are given below. You can make up your own. You could also ask each pupil to write one question, and then give all the questions orally in class.

1	What is the diameter of a tenpin bowl?	21.6 cm
2	How much is a golf ball?	£1.30
3	Which is larger – a table tennis or golf ball?	Golf ball
4	What is the weight of a table tennis ball?	2.5 g
5	What is the cost of two men's shots?	£20
6	What is the weight of a woman's shot in kg?	4 kg
7	How much is the cheapest football?	£7.50
8	How much is the dearest football?	£30
9	How much for a pair of tenpin bowls?	£80
10	How much might you expect for one tenpin bowl? (Though it might not be possible to buy one!)	£40
11	Which is the smallest ball?	Table tennis
12	Which ball has the greatest diameter?	Football
13	Which is the lightest ball?	Table tennis
14	Which is the heaviest ball?	Men's shot
15	Which balls could weigh 160 grams?	Cricket, pool
16	Which balls could have a diameter of 5.7 cm?	Rounders, Pool
17	What is the smallest diameter of a football?	21.8 cm
18	What is the largest diameter of a rounders ball?	6.0 cm

19	What is the diameter in mm of a tennis ball?	68 mm
20	How much might one tennis ball cost?	£1.30
21	How much might one table tennis ball cost?	50p
22	How many rounders balls could you buy for £20?	5
23	How many golf balls could you buy for £5?	3
24	How many cricket balls could you buy for £50?	6
25	How much for nine tennis balls?	£11.70
26	How much for eight bowls and two jacks?	£224
27	Give two balls that weigh roughly the same.	Men's shot, tenpin or cricket, pool or football, netball
28	How much heavier is the men's shot than the women's shot?	3260 g
29	How many cricket balls weigh just under 1 kg?	6
30	What is the weight in kilograms of a tenpin bowl?	7.26 kg

33 Inputs and outputs

7C/27

This unit introduces pupils to equivalent expressions such as $3(a + 6) = 3a + 18$ and $2(a - 4) = 2a - 8$.

In this unit, arrow diagrams are drawn with circles and ellipses. Pupils may find it easier to use squares and rectangles in their diagrams.

p 240 **A** Input and output machines	Using number machines
p 242 **B** Shorthand rules	Using algebra to write rules such as $n \to 5n - 3$, $n \to 5(n - 3)$
p 244 **C** Evaluating expressions	Substituting into expressions such as $2(n + 5)$, $2n + 5$
p 245 **D** Different rules?	Discussing equivalent rules such as $n \to 5n - 6$, $n \to 5(n - 3)$
p 246 **E** Equivalent expressions	Multiplying out brackets such as $6(n + 2) = 6n + 12$
p 247 **F** Number tricks	Using algebra to explain 'think of a number' puzzles

Essential

Sheets 148 and 149

Practice booklet pages 76 to 78

A Input and output diagrams (p 240)

A5, A6 These questions should get pupils thinking about looking for shorter chains. By observing the number patterns (it may be helpful to add some consecutive values to the table) they should be able to find shorter chains.

At the end of the unit, pupils could come back to these and try to prove their results using algebra.

B Shorthand rules (p 242)

Pupils have met the fact that we can write $a \times 4$ as $4a$. Here the notation is slightly extended, in that we write $(c + 5) \times 4$ as $4(c + 5)$. Revision of 'do what is in the brackets first' may be needed with some pupils.

The emphasis here is on using letters to write rules. However, pupils consider numerical outputs in question B5.

T

'Supplemented with many more examples of × then + or + then × .'

◊ In discussing the notation, bring out the fact that any letter can be used to represent an input number, and that when multiplication is involved, the number comes first and the multiplication sign is omitted.

Mention that because 4 × 3 = 3 × 4 we can (and do!) always put the number before the letter. You may want to use several examples like those in the introduction to ensure that pupils are secure with the notation. Include the fact that we can write $s \times 1$ as simply s.

◊ Ask pupils to choose some input numbers and find the outputs for the two diagrams at the top of page 242. They could think about whether for each input number the corresponding outputs are always different, and try this for different pairs of diagrams where the operations are swapped round.

C Evaluating expressions (p 244)

> Sheet 148

T

◊ Your discussion should include how to use arrow diagrams to evaluate expressions with and without brackets. You could swap the operations in the example on the page to consider $5(p - 3)$.

◊ To break the ground for the 'Cover up' game, pupils could discuss different possible rules for a single input–output pair. For example, ask pupils in pairs to find as many rules as they can that fit $2 \rightarrow 5$.

D Different rules? (p 245)

T

This teacher-led section is for pupils to begin to consider when two rules are equivalent by looking at numerical examples.

◊ Pupils begin by matching each rule with an arrow diagram as they did in section C.

Now ask pupils to match each number pair in the bubble to as many of the rules as they can. They could record their results in tables:

$n \rightarrow 2n + 10$	$n \rightarrow n + 10$	$n \rightarrow n + 1 + 9$	$n \rightarrow 2(n + 5)$
$10 \rightarrow 30$	$1 \rightarrow 11$	$1 \rightarrow 11$	$10 \rightarrow 30$
$3 \rightarrow 16$			$3 \rightarrow 16$
$2 \rightarrow 12$	$2 \rightarrow 12$	$2 \rightarrow 12$	$2 \rightarrow 12$ ...

Pupils should notice that the lists for $n \rightarrow 2n + 10$ and $n \rightarrow 2(n + 5)$ are identical, as are the lists for the other pair of rules.

They are likely to be able to see that **any** input–output pair that fits $n \to n + 10$ will fit $n \to n + 1 + 9$ as the equivalence of 'add 10' and 'add 1 and then add 9' is familiar to them.

However, it is less obvious that **any** number pair that fits $n \to 2n + 10$ will fit $n \to 2(n + 5)$. The question of whether or not the two rules really are equivalent can be left unresolved till section E.

◊ In the second box, pupils copy and complete the table for each rule, perhaps drawing an arrow diagram for each rule first.

They should find that the rules seem to match up to give two equivalent pairs. Again the question of whether or not we really do have pairs of equivalent rules can be left unresolved till section E. Some pupils may be able to appreciate that agreement for some numerical examples does not in general give conclusive proof of equivalence.

E Equivalent expressions (p 246)

This section follows on from the teacher-led section D. The initial discussion should include examples that involve subtraction.

> Sheet 149 (one for each player)

◊ Diagrams such as the one on page 246 can help to explain why, for example, $3(n + 2) = (3 \times n) + (3 \times 2) = 3n + 6$.

You could also use simple areas to show the equivalence. For example:

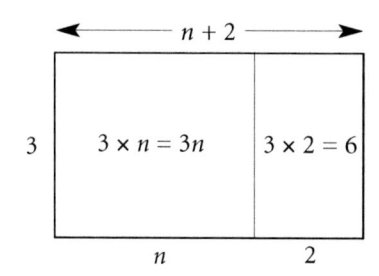

[T]
'Used Cuisenaire blocks and colour trains to find different ways of expressing the same length.
For example:

| r | b | r | b |

is $2(r + b)$ and

| r | r | b | b |

is $2r + 2b$.'

Include some examples where the number comes first in the brackets, for example $3(2 + n) = 6 + 3n$ or $3n + 6$.

◊ Discussion of numerical examples can help to explain the equivalence when the expression involves subtraction.

You could start by looking at addition and noticing, for example,
$3 \times (100 + 2) = (3 \times 100) + (3 \times 2) = 300 + 6 = 306$

Extend to subtraction:
$3 \times (100 - 2) = (3 \times 100) - (3 \times 2) = 300 - 6 = 294$

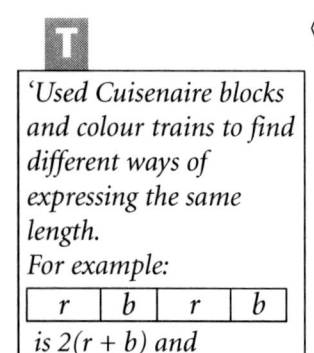

'To avoid overenthusiastic shouting, I asked players to put their hand on a pile if there was a snap.'

Expression snap

◊ You may wish to copy the sets of cards on to different coloured card. If pairs of pupils have different coloured sets, it will help them sort the cards after each game.

F Number tricks (p 247)

This extension material gives pupils the opportunity to use what they have learned to explain and create number puzzles where you always end up with the number first thought of.

A Input and output machines (p 240)

A1 (a) 3 (b) 12

A2 (a) 13 (b) 16 (c) 1 (d) 37

A3 (a) $17 \xrightarrow{\times 2} 34 \xrightarrow{-5} 29$

(b) $6 \xrightarrow{\times 3} 18 \xrightarrow{-4} 14$

(c) $12 \xrightarrow{+1} 13 \xrightarrow{\times 2} 26$

(d) $2 \xrightarrow{\times 8} 16 \xrightarrow{+10} 26$

A4 (a) 2 → **5**
4 → **11**
7 → 20
10 → 29

(b) 6 → **7**
16 → **12**
0 → 4
3 → $5\tfrac{1}{2}$

(c) 9 → **21**
4 → **11**
5 → 13
7 → 17

(d) 1 → **11**
5 → **31**
6 → 36
$1\tfrac{1}{2}$ → $13\tfrac{1}{2}$

*__A5__ (a) 2 → **6**
8 → **24**
10 → **30**
5 → 15
33 → 99
0.5 → 1.5

(b) 1 → **2**
5 → **6**
39 → **40**
9 → 10
$7\tfrac{1}{2}$ → $8\tfrac{1}{2}$
99 → 100

*__A6__ (a) ◯ $\xrightarrow{\times 3}$ ◯

(b) ◯ $\xrightarrow{+1}$ ◯

B Shorthand rules (p 242)

B1 (a) $b \xrightarrow{\times 3} 3b \xrightarrow{-2} 3b-2$

(b) $b \to 3b - 2$

B2 (a) $p \xrightarrow{\times 3} 3p \xrightarrow{+4} 3p+4$

$p \to 3p + 4$

(b) $p \xrightarrow{+4} p+4 \xrightarrow{\times 3} 3(p+4)$

$p \to 3(p + 4)$

B3 Diagram C is correct.

$s \xrightarrow{+2} s+2 \xrightarrow{\times 3} 3(s+2)$

B4 (a) $a \xrightarrow{\times 5} 5a \xrightarrow{-3} 5a-3$

$a \to 5a - 3$

(b) $a \xrightarrow{-3} a-3 \xrightarrow{\times 5} 5(a-3)$

$a \to 5(a - 3)$

(c) $w \xrightarrow{+4} w+4 \xrightarrow{\times 7} 7(w+4)$

$w \to 7(w + 4)$

(d) $w \xrightarrow{\times 7} 7w \xrightarrow{+4} 7w+4$

$w \to 7w + 4$

33 Inputs and outputs • 185

B5 (a) s →×2→ 2s →+1→ 2s + 1

(b) t →−9→ t − 9 →×3→ 3(t − 9)

(c) w →×5→ 5w →−7→ 5w − 7

(d) x →+7→ x + 7 →×6→ 6(x + 7)

(e) y →×7→ 7y →+3→ 7y + 3

(f) z →+5→ z + 5 →×2→ 2(z + 5)

When the input is 100, the outputs are
(a) 201 (b) 273 (c) 493
(d) 642 (e) 703 (f) 210

B6 (a) d →+4→ d + 4 →×3→ 3(d + 4)
d → 3(d − 4)

(b) s →×2→ 2s →+3→ 2s + 3
s → 3s + 3

(c) g →+1→ g + 1 →×3→ 3(g + 1)
g → 3(g + 1)

(d) a →×3→ 3a →+1→ 3a + 1
a → 3a + 1

(e) e →×1→ e →+10→ e + 10
e → e + 10

(f) j →−3→ j − 3 →×2→ 2(j − 3)
j → 2(j − 3)

C Evaluating expressions (p 244)

C1 16

C2 (a) 9 (b) 8 (c) 26 (d) 6

C3 (a) 28 (b) 32 (c) 36 (d) 60

C4 (a) 2 → **13** (b) 3 → **17**
(c) 4 → **21** (d) 10 → **45**

C5 (a) 1 → 4
(b) 1 → 4 and 2 → 9
(c) 1 → 4 and 5 → 12
(d) 5 → 0
(e) 2 → 9
(f) 3 → 0 and 4 → 5

C6 The pupil's three rules for 1 → 6

Cover up
One solution to board A is

p → 5p − 6	a → 5(a −2)
t → 3t	b → 3b + 4	m → m + 6
	x → 2x + 8	
w → 3w − 2	y → 3(y + 2)

A solution to board B is

w → 3w − 2	p → 5p − 6	b → 3b + 4
y → 3(y +2)	a → 5(a − 2)	
		m → m + 6
t → 3t	x → 2x + 8

E Equivalent expressions (p 246)

E1 2(x + 4) and 2x + 8
2(x + 8) and 2x + 16
2(x + 2) and 2x + 4
The odd one is 2(x + 16).

E2 3a + 18 and 3(a + 6)
3(a − 2) and 3a − 6
3a − 18 and 3(a − 6)
The odd one is 3a − 2.

E3 (a) $3x + 15$ (b) $7b + 21$
(c) $5c + 20$ (d) $4d + 4$
(e) $12e + 24$ (f) $3w - 6$
(g) $4a - 12$ (h) $70m - 70$
(i) $12 + 6f$ or $6f + 12$
(j) $3y - 27$
(k) $40 + 8k$ or $8k + 40$
(l) $6n - 12$

E4 (a) $2a + 10 = \mathbf{2}(a + 5)$
(b) $2(a - 6) = 2a - \mathbf{12}$
(c) $4a - 12 = 4(a - \mathbf{3})$
(d) $3(a + 7) = \mathbf{3}a + 21$
(e) $5a - 20 = \mathbf{5}(a - \mathbf{4})$
(f) $\mathbf{7}a + 14 = 7(a + \mathbf{2})$
(g) $2(\mathbf{6} + \mathbf{p}) = 12 + 2p$ or $2(\mathbf{p} + \mathbf{6}) = 12 + 2p$
(h) $4a - 32 = \mathbf{4}(\mathbf{a} - \mathbf{8})$

F Number tricks (p 247)

*F1 (b) You get the number you started with each time.
(c) $n \to n + 2 \to 5(n + 2) = 5n + 10 \to 5n \to n$
(d) The pupil's explanation, for example: 'You start with n and end with n'.

*F2 Add **8**.   Divide by **8**.

*F3 The pupil's instructions

What progress have you made? (p 248)

1 $5 \xrightarrow{+4} 9 \xrightarrow{\times 3} 27$

2 $3 \to \mathbf{11}$
   $10 \to \mathbf{46}$
   $5 \to \mathbf{21}$
   $8 \to \mathbf{36}$

3 (a) $a \xrightarrow{\times 2} 2a \xrightarrow{-5} 2a - 5$
   $a \to \mathbf{2a - 5}$

(b) $s \xrightarrow{+3} s + 3 \xrightarrow{\times 4} 4(s + 3)$
   $s \to \mathbf{4(s + 3)}$

4 $y \xrightarrow{\times 3} 3y \xrightarrow{+6} 3y + 6$

5 (a) 19   (b) 27   (c) 18   (d) 28

6 (a) $6 \to \mathbf{15}$ (b) $1 \to \mathbf{0}$

7 (a) $4x - 12$ (b) $5s + 5$
  (c) $2b + 18$ (d) $6k - 24$

Practice booklet

Sections A and B (p 76)

1 (a) $9 \xrightarrow{\times 2} 18 \xrightarrow{-12} 6$

(b) $1 \xrightarrow{+5} 6 \xrightarrow{\times 5} 30$

2 (a) $1 \to \mathbf{1}$ (b) $2 \to \mathbf{45}$
     $3 \to \mathbf{9}$ $4 \to \mathbf{63}$
     $\mathbf{4} \to 13$ $\mathbf{1} \to 36$
     $\mathbf{10} \to 37$ $\mathbf{7} \to 90$

3 (a) $n \xrightarrow{\times 3} 3n \xrightarrow{+1} 3n + 1$
   $n \to 3n + 1$

(b) $n \xrightarrow{+5} n + 5 \xrightarrow{\times 2} 2(n + 5)$
   $n \to 2(n + 5)$

(c) $n \xrightarrow{\times 2} 2n \xrightarrow{-3} 2n - 3$
   $n \to 2n - 3$

(d) $n \xrightarrow{-4} n - 4 \xrightarrow{\times 5} 5(n - 4)$
   $n \to 5(n - 4)$

(e) $n \xrightarrow{+1} n + 1 \xrightarrow{\times 10} 10(n + 1)$
   $n \to 10(n + 1)$

(f) $n \xrightarrow{\times 7} 7n \xrightarrow{-6} 7n - 6$
   $n \to 7n - 6$

4 (a) $p \xrightarrow{\times 2} 2p \xrightarrow{+3} 2p+3$

(b) $m \xrightarrow{\times 2} 2m \xrightarrow{+6} 2m+6$

(c) $x \xrightarrow{-3} x-3 \xrightarrow{\times 2} 2(x-3)$

(d) $b \xrightarrow{\times 3} 3b \xrightarrow{+10} 3b+10$

(e) $t \xrightarrow{+4} t+4 \xrightarrow{\times 3} 3(t+4)$

(f) $n \xrightarrow{-1} n-1 \xrightarrow{\times 5} 5(n-1)$

When the input is 10, the outputs are

(a) 23 (b) 26 (c) 14
(d) 40 (e) 42 (f) 45

Section C (p 77)

1 (a) 10 (b) 22 (c) 2 (d) 12

2 (a) 5 (b) 35 (c) 119 (d) 17

3 (a) $1 \to \mathbf{18}$ (b) $3 \to \mathbf{24}$
(c) $10 \to \mathbf{45}$ (d) $100 \to \mathbf{315}$

4 $3(m+1)$ and $5(m+1)$ do not have the value 10 when $m = 3$.

5 (a) 49 (b) 37 (c) 3
(d) 11 (e) 20 (f) 24

6 (a) 4 (b) 1 (c) 5
(d) 1

Section E (p 78)

1 $2(x+4) = 2x+8$
$2(x+2) = 2x+4$
$4(x+2) = 4x+8$
$4(x+1) = 4x+4$

2 $3(x-6) = 3x-18$
$3(x-2) = 3x-6$
$3(x-18) = 3x-54$
$3(x-1) = 3x-3$
Odd one out: $3x-2$

3 (a) $2a+6$ (b) $5x-5$
(c) $4p+40$ (d) $10p+40$
(e) $7y+21$ (f) $100c-100$
(g) $250+25t$ or $25t+250$
(h) $3n-30$

4 (a) $2(x+3) = 2x+\mathbf{6}$
(b) $\mathbf{4}(x+4) = 4x+16$
(c) $5x+10 = \mathbf{5}(x+\mathbf{2})$
(d) $2x-\mathbf{6} = 2(x-3)$
(e) $3x+\mathbf{6} = \mathbf{3}(x+2)$
(f) $\mathbf{4}(x+6) = 4x+24$

*5 (a) $8 \to \mathbf{29}$
$2 \to \mathbf{5}$
$6 \to 21$
$10 \to 37$

(b) The chain can be replaced with

$\bigcirc \xrightarrow{\times 4} \bigcirc \xrightarrow{-3} \bigcirc$

34 Action and result puzzles (p 249) 7T/32, 7C/9

In each puzzle, the action cards show operations to be performed on a starting number and the result cards show the results. Pupils match up the results with the actions.

The puzzles provide number practice and an opportunity to apply some logical thinking. They may reveal misconceptions about number.

Essential	Optional
Puzzles on sheets 106 and 107	Sheet 110 (blank cards)
Scissors	Sheets 108 (harder) and 104 (easier)
	Transparencies of some sheets, cut into puzzle cards

T

'I put copies of the games into labelled boxes. I explained what each game was about and allowed them to choose their own games.'

A selection from the puzzles should be used (listed below roughly in order of difficulty).

Sheet 106   327 puzzle (+ and –, three-digit numbers)
            6.5 puzzle (+ and –, includes simple fractions and decimals)
            3679 puzzle (+ and –, four-digit numbers)

Sheet 107   36 puzzle (+, –, × and ÷, includes simple decimals and negative numbers)
            q puzzle ('logic' puzzle, + and –, two-digit numbers)
            h puzzle ('logic' puzzle, +, –, × and ÷, two-digit numbers)

Pupils who do well with these can do the following as extension.

Sheet 108   16.5 puzzle (+, –, × and ÷, decimals and negative numbers)
            s puzzle ('logic' puzzle, the number s has to be found)
            12 puzzle (a, b and c are numbers to be found)

Any pupils who need a gentler start could do the following.

Sheet 104   44 puzzle (+ and –, whole numbers ≤ 100)
            60 puzzle (+, –, × and ÷, simple 2- and 3-digit numbers)
            5.5 puzzle (+ and –, decimals with 0.5 only)

◊ This has worked well with pupils sitting in pairs on tables of four. When each pair had matched the cards, all four pupils discussed what they had done. An aim is to encourage mental number work. However, pupils may want to do some calculations and demonstrate things to their group using pencil and paper. It is not intended that a calculator should be used.

'I copied the cards on to pieces of acetate which could be moved about on the OHP. Pupils went to the OHP to show how the cards matched up.'

◊ Puzzles that pupils find easy can be done without cutting out the cards: they simply key each action card to its result card by marking both with the same letter. However, something may be learnt from moving cards around to try ideas out before reaching a final pairing, and some puzzles are almost impossible unless they are done this way.

◊ Solutions can be recorded by
- keying cards to one another with letters as described above
- sticking pairs of cards on sheets or in exercise books
- writing appropriate statements, such as 8 – 3 = 5

◊ After pupils have solved some puzzles, they can make up some of their own (using the blank cards) to try on a partner. This may tell you something about the limits of the mathematics they feel confident with. Some should be able to make up puzzles of the q and h type.

Sheet 106

327 puzzle

Action	Result
– 30	297
+ 40	367
– 110	217
+ 700	1027
+ 390	717
– 89	238
+ 651	978
– 207	120

3679 puzzle

Action	Result
+ 30	3709
– 2030	1649
+ 45	3724
– 95	3584
– 2300	1379
+ 4205	7884
+ 999	4678
– 680	2999

Sheet 107

36 puzzle

Action	Result
÷ 9	4
– 40	⁻4
× 3	108
+ 27	63
÷ 8	4.5
× 1.5	54
+ ⁻50	⁻14
÷ 24	1.5

q puzzle: $q = 27$

h puzzle: $h = 9$

6.5 puzzle

Action	Result
+ 2.5	9
– 3.5	3
+ 9	15.5
+ 2.25	8.75
– 0.5	6
+ 4.75	11.25
– 1½	5
– ¾	5.75

Sheet 108 (extension)

16.5 puzzle

Action	Result
+ 3.5	20
÷ 10	1.65
– 17	⁻0.5
× ⁻2	⁻33
÷ 0.5	33
+ ⁻3.5	13
– 6.05	10.45
× 0.1	1.65

s puzzle: $s = 4.5$

12 puzzle: $a = 0.5$, $b = 4$ and $c = 2$
(or $b = 2$ and $c = 4$), $d = 10$

Sheet 104 (gentler start)

44 puzzle

Action	Result
– 43	1
+ 43	87
– 29	15
+ 29	73
+ 50	94
– 9	35
+ 56	100
– 14	30

5.5 puzzle

Action	Result
+ 0.5	6
– 0.5	5
+ 2.5	8
– 2.5	3
– 1.5	4
+ 9	14.5
+ 3.5	9
– 3.5	2

60 puzzle

Action	Result
÷ 10	6
– 10	50
– 12	48
÷ 5	12
× 3	180
× 10	600
+ 100	160
÷ 3	20

190 • 34 Action and result puzzles

35 Perpendicular and parallel lines

7T/20, 7T/22, 7C/24, 7C/30

In several places here, letters are used to label points, then line segments or polygons are referred to in terms of those letters. Some pupils find this convention difficult, but it is an indispensable one and it is important to help them feel comfortable with it.

p 250 **A** Right angles

p 253 **B** The shortest route to a line

p 254 **C** Parallel lines

Essential

Set square
Squared paper

Practice booklet pages 79 to 80

A Right angles (p 250)

For some pupils, this section will be revision.

Set square

A2 Some pupils may be uncertain about the points of the compass and find it difficult to relate them to turning; clockwise and anticlockwise may also be a problem. If so, it's worth developing this question into a class activity. Establish what direction is north in relation to your classroom. Then pupils take turns to stand facing a given direction, follow instructions to turn clockwise or anticlockwise through a right angle and then say what direction they are now facing.

A4 You can develop this question and A5 into class discussion if pupils seem likely to benefit from it.

'Good practice with using ruler and set square.'

A6 An aim here is to develop accuracy. You can introduce the idea of measuring both diagonals of a rectangle to check its 'squareness'.

B The shortest route to a line (p 253)

> Set square, squared paper

'NONE of the class realised that the shortest line was the perpendicular and I had to class-teach it.'

◊ Many pupils find it difficult to see that the shortest path from a point to a line is perpendicular to the line. When they have drawn point H and the line AB (the exact positions don't matter), you can ask them to draw the shortest route; they can then challenge their neighbour to draw a shorter route than theirs. The angle each drawn route makes with AB can be measured. This may lead to further drawing and angle measuring.

◊ A complementary approach is to draw this on a transparency and display it on the OHP. Ask how to draw the line from H to line AB. Most pupils should have no difficulty in saying 'straight down', so you can draw the path, point to one of the angles it makes with AB and ask the size of the angle. The answer 'a right angle' should be forthcoming. Check that everybody agrees.

You can then turn the acetate round like this and ask what the angle is now. If any pupils hesitate, move the acetate back so AB is 'horizontal' again and check they can say what the angle is. Move AB back to its oblique position and get them to tell you whether anything has happened to change the angle between the path and AB, and so on ...

C Parallel lines (p 254)

> Squared paper

◊ Some pupils will have met the word 'parallel' before, so you could begin by asking someone to draw a pair of parallel lines on the board and asking the rest of the class what it is that makes the lines parallel. Ideas such as 'they never meet', 'they go in the same direction' and 'they stay the same distance apart' may arise. Clarify that parallel lines do not have to be the same length or lined up in some special way. At the same time you can ask about where parallel lines occur in the real world.

The word 'parallel' is restricted to straight lines here, though it is used in everyday language to refer to curved lines (rails on a railway, lines of latitude) that never meet.

C3 To identify a set of parallel lines, some pupils may need help to see that that lines are all made up of segments that go, for example, one square across and two squares up.

A Right angles (p 250)

A1 90°

A2 (a) North (b) North-east
 (c) North-east

A3 (a) The angles between them (56° and 34°) add up to 90°.
 (b) *a* and *h*, *b* and *d*

A4 2 o'clock

A5 Ten to nine (8:50)

A6 The pupil's drawing of the rectangle

A7 The pupil's drawing (the fourth side should be 11 cm long)

A8 The pupil's drawing (the missing sides should be 2.7 cm and 7.2 cm long)

A9 The pupil's drawing

A10 *b* and *f*

A11 *q* and *v*

A12 *d* and *e*, *f* and *i*, *g* and *h*

B The shortest route to a line (p 253)

B1 (a) [diagram]
 (b) The swimmer should swim to edge ED.

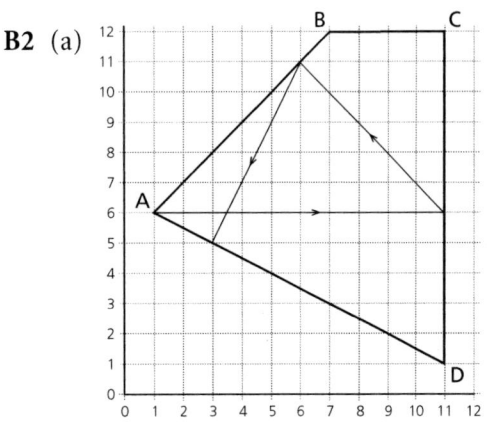

B2 (a) [diagram]
 (b) She gets to the point (3, 5).

C Parallel lines (p 254)

C1 They are parallel. In each case, for every 3 squares to the right the line goes 2 squares down.

C2 The pupil's drawing; the first line should go through P and (4, 3) and the second through Q and (6, 9).

C3 *a*, *b*, *h* and *j*; *c*, *i* and *k*; *d*, *f* and *g*; *e* and *l*

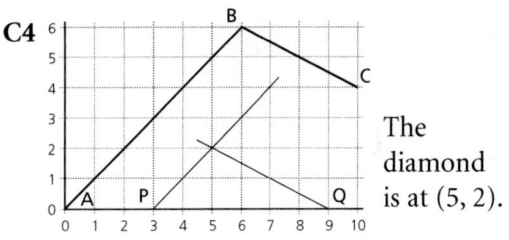

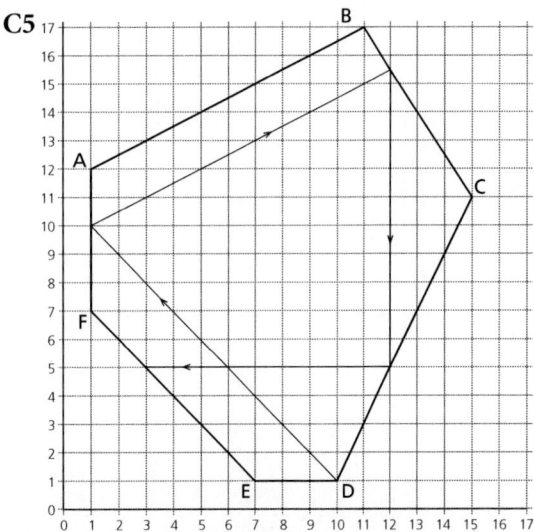

C4 [diagram] The diamond is at (5, 2).

C5 [diagram]
The treasure is buried at (3, 5).

35 Perpendicular and parallel lines • 193

C6 The two lengths measured on each transversal are the same.
If you mark points 12 cm apart and 4 cm apart, the lengths measured on each transversal are in the ratio 2 : 1.
In general, if the distances between the dots on the parallel lines are in the ratio $n : 1$, the lengths measured on each transversal are in the ratio $n - 1 : 1$.

C7 (a) Opposite sides are equal.
Opposite angles are equal.
(b) It has rotation symmetry of order 2.
(c) It does not have reflection symmetry (unless a rhombus happens to have been drawn).
(d) A parallelogram

C8 (a) All four sides are equal.
(b) Order 2
(c) Each of the diagonals is a line of reflection symmetry.
(d) A rhombus

C9 (a) Opposite sides are equal.
(b) All four angles are right angles.
(c) It has two lines of symmetry going through its centre, parallel to the sides.
(d) A rectangle

What progress have you made? (p 258)

1 a and b, c and d, e and f

2 The pupil's drawing

3 The pupil's drawing of two parallel lines 4 cm apart.

4 (a) and (b)

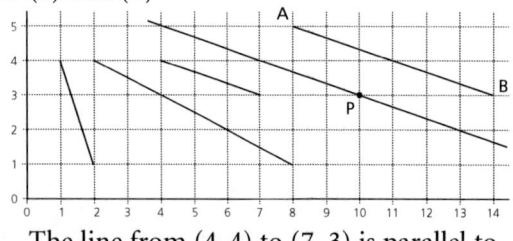

The line from (4, 4) to (7, 3) is parallel to AB.

Practice booklet

Sections A and B (p 79)

1 (a) South   (b) South-east

2 a and d, b and h, c and f

3 (a), (b) and (c)

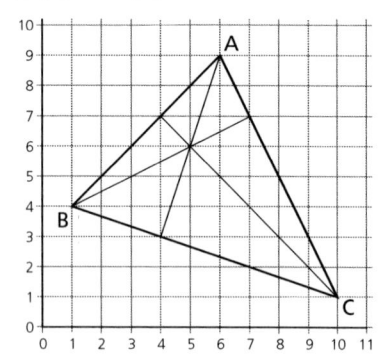

(d) All three pass through one point.

4 (a) and (b)

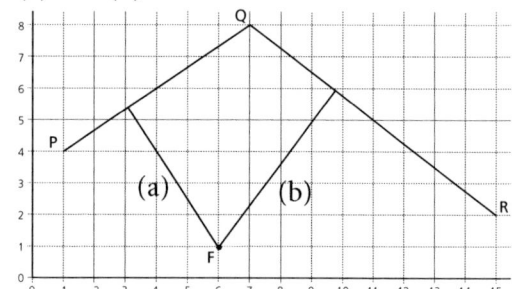

Section C (p 80)

1 a and c, b and e, d and g, f and h

2 (a) C   (b) E   (c) G

3 (3, 4)

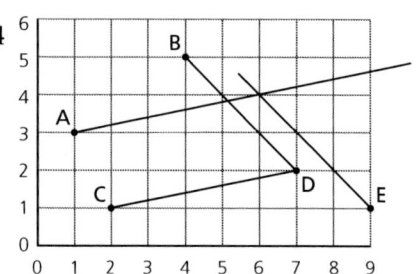

4

The cargo hold is at (6, 4).

Review 4 (p 259)

Essential
Set square

1 C: $n \to 3n + 1$

2 (a) $\frac{6}{8}$ or $\frac{3}{4}$ (b) $\frac{2}{6}$ or $\frac{1}{3}$ (c) $\frac{2}{4}$ or $\frac{1}{2}$

3 (a) Across '+ 5', down '+ 1'
   (b) Across '+ 2', down '− 3'

4 (a) $9 \xrightarrow{\times 2} 18 \xrightarrow{+5} 23$

   (b) $2 \xrightarrow{\times 9} 18 \xrightarrow{+14} 32$

5 $\frac{1}{2}$ and 0.5, 0.4 and $\frac{2}{5}$, 0.8 and $\frac{4}{5}$, 0.2 and $\frac{1}{5}$

6 $\frac{3}{5}$

7 (a)
$n \to 2n - 3$
$6 \to 9$
$10 \to 17$
$1\frac{1}{2} \to 0$
$1 \to {}^-1$

   (b)
$n \to \frac{n}{5}$
$15 \to 3$
$50 \to 10$
$10 \to 2$
$0 \to 0$

   (c)
$n \to 11 - n$
$2 \to 9$
$4 \to 7$
$8 \to 3$
$11 \to 0$

   (d)
$n \to n + 7$
$8 \to 15$
$2 \to 9$
$10 \to 17$
$12 \to 19$

8 $\frac{1}{4}, \frac{37}{100}, \frac{7}{10}$

9 (a) and (b)
The coordinates of F are (12, 4).

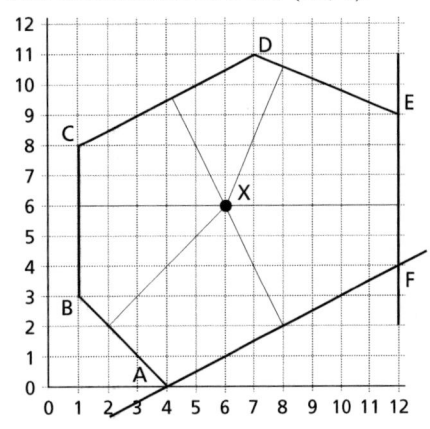

   (c) The route to CD is the shortest.

10 (a)
$+3 \to$

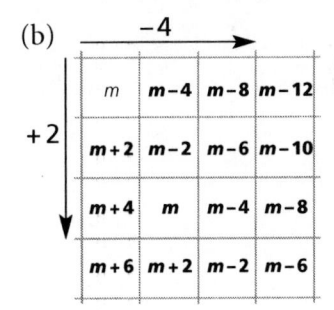

	n	$n+3$	$n+6$
$-1 \downarrow$	$n-1$	$n+2$	$n+5$
	$n-2$	$n+1$	$n+4$

   (b)
$-4 \to$

	m	$m-4$	$m-8$	$m-12$
$+2 \downarrow$	$m+2$	$m-2$	$m-6$	$m-10$
	$m+4$	m	$m-4$	$m-8$
	$m+6$	$m+2$	$m-2$	$m-6$

11 (a) $\frac{31}{100}$ (b) $\frac{6}{10}$ or $\frac{3}{5}$

12 (a) $h + 7$ (b) $k + 3$ (c) $m - 7$
   (d) $2w + 4$ (e) $2n - 5$ (f) $3x - 3$

13 (a) $\frac{2}{8}$ or $\frac{1}{4}$ (b) $\frac{4}{8}$ or $\frac{1}{2}$

14 (a) $s \xrightarrow{+2} (s+2) \xrightarrow{\times 4} 4(s+2)$

   (b) $z \xrightarrow{\times 5} 5z \xrightarrow{+6} 5z+6$

(c) w → w−9 → 3(w−9) [−9, ×3]

(d) p → 4p → 4p−7 [×4, −7]

(e) h → h+9 → 3(h+9) [+9, ×3]

15 (a) 48 (b) 50.5 (c) 4 (d) 3

16 (a) 9.6 (b) 3.75

17 (a) 4p + 8 (b) 3x − 21
(c) 5d + 25 (d) 14 + 7t

18 $\frac{5}{100}$, 0.09, 0.3, 0.63, $\frac{3}{4}$

19 (a) 2a + 8 = 2(a + **4**)
(b) **9**b − 45 = 9(**b** − 5)
(c) 7(s − 2) = 7s − **14**

(There are other possible answers for (b).)

Mixed questions 4 (Practice booklet p 82)

1 (a)

n → 4n − 3
1 → 1
11 → 41
5 → 17
6 → 21

(b)

n → $\frac{n}{4}$
16 → 4
8 → 2
28 → 7

(c)

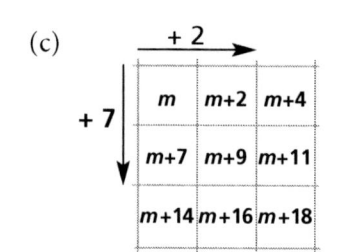

n → 5n
1 → 5
10 → 50
4 → 20
6 → 30

2 $\frac{3}{6}$ or $\frac{1}{2}$

3 (a) [−5 down, −2 right]

	−2		
−5	20	18	16
	15	13	11
	10	8	6

(b) [+6 down, −4 right]

	−4		
+6	11	7	3
	17	13	9
	23	19	15

(c) [+2 right, +7 down]

	m	m+2	m+4
+7	m+7	m+9	m+11
	m+14	m+16	m+18

4 (a) 12 (b) 10 (c) 66 (d) 59.5

5 $\frac{1}{25}$

6 (a) n + 11 (b) n − 1
(c) 2n + 4 (d) 3n − 5

7 (a) 0.3 (b) 0.8 (c) 2.75
(d) 6.01 (e) 4.1

8 (a) 3(p + 5) = 3p + **15**
(b) 4q − 12 = **4**(q − **3**)
(c) 5(r − **3**) = 5r − 15
(d) **7**s + 42 = **7**(s + 6)

9 (a) $\frac{6}{8}$ or $\frac{3}{4}$ (b) $\frac{4}{8}$ or $\frac{1}{2}$
(c) $\frac{4}{8}$ or $\frac{1}{2}$ (d) $\frac{2}{8}$ or $\frac{1}{4}$
(e) $\frac{1}{8}$ (f) 0

10 $\frac{2}{100}$, $\frac{3}{20}$, 0.19, $\frac{1}{5}$, 0.78, $\frac{8}{10}$

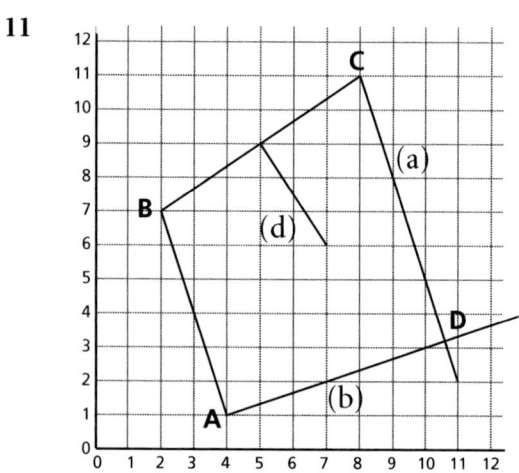

11

(c) Shape ABCD has two right angles.

196 • *Mixed questions 4*

36 Practical problems

7T/34

These use equipment so need to be done in class. Pupils need not do all the tasks: just doing some will tell you a lot about how well they can measure, estimate and apply number skills in problem solving.

One approach is to set up a 'circus' of tables; on each table is the equipment for one task and a label giving its name and page number (tasks with readily available equipment can be duplicated on more than one table). Pupils move round the circus following the instructions in the pupil's book. It is a good idea to have some 'exercise' work ready in case there is a log-jam as pupils go around the circus.

Alternatively, you can set up just one task (possibly in duplicate), and while the rest of the class get on with written work individuals or small groups come out in turn to do it.

'Weighty problems' is for a small group of pupils; the others can be done individually or in pairs.

Many pupils find weighing difficult, whether interpreting scale graduations on mechanical scales or coping with decimal places on digital balances. So it may be a good idea to add some straightforward weighing to the collection of tasks.

Weighty problems (p 262)

> Scales or electronic balance
> Two stones, say about 4 cm and 8 cm in diameter
> A collection of familiar objects, including one with its weight clearly marked on it (for example a 500 g or 1 kg bag of sugar)

◊ In task 1 each group's estimates could be displayed on a dot plot (see 'Comparisons') and ideas of spread and over- or under-estimation discussed by the class. It is a good idea if everybody in the group checks the reading on the scales when the stones are weighed.

◊ After task 2 pupils could discuss ways of deciding which estimated order of weights was the best.

A related activity that goes well is for a pupil to hold the object of known weight (bag of sugar or whatever) in one hand and a different object in the other; the pupil estimates the weight of the other object then checks by weighing.

Beans (p 262)

> Two identical large sweet jars with lids, one empty and the other at least half filled with dried beans or pasta shapes with its lid taped down (butter beans are suitable, but not red kidney beans or other varieties that are poisonous when uncooked)
> About 100 extra beans or pasta shapes of the type in the jar
> An electronic balance or scales sensitive enough to weigh a few grams

◊ These are two approaches pupils have used to start solving the problem.
- Putting a layer of the extra beans into the empty jar and measuring the layer's height.
- Finding the weight of the beans in the jar by subtracting the weight of the empty jar.

In the second case, some go on to weigh a single bean. If so, ask them to check whether the beans have the same or different weights. If they have different weights, can pupils suggest a way to deal with this?

Cornflakes (p 263)

> A full box of cornflakes, with its price
> A cereal bowl
> An electronic balance or scales

◊ Some pupils may need a hint to work out the weight of the cornflakes without the bowl.

◊ You could extend the work to comparing the cost of different types of cereal or comparing the cost of a serving-sized box with that of the same amount of cereal in a full-sized box.

Getting better (p 263)

> A 5 ml spoon labelled 5 ml
> A container with a scale graduated in ml
> Three different sized medicine bottles (one less than 60 ml) distinguished by colour or labelling, but without their capacities marked
> Water, a tray and some paper towels

◊ A prepared answer sheet may help weaker pupils.

◊ Follow-up might include
- discussion of the appropriate level of accuracy
- estimation of capacity from knowing that 1 cm cube has a volume of a millilitre and a 10 cm cube has a volume of a litre
- estimation by pupils, perhaps as a homework assignment, of their daily fluid intake

37 Comparisons   7C/25

This unit introduces median and range and uses these to compare sets of data. Pupils generate their own data in a reaction time experiment and use it to make comparisons. Other data handling projects are suggested and there is advice on writing up.

T	p 264 **A** Comparing heights	
	p 266 **B** Median	
	p 268 **C** Range	
T	p 270 **D** How fast do you react?	Making and using a simple reaction timer to generate data for comparison
	p 271 **E** Summarising data	Using extremes, median and range as a summary
T	p 272 **F** Writing a report	Discussing a specimen report
T	p 274 **G** Projects	Activities for the class, and for pairs or groups, generating data for comparison

Essential
Sheet 141
Practice booklet pages 84 and 85

Optional
Sheets 142 to 145

A Comparing heights (p 264)

T

◊ The discussion here is intended to be open, with no particular method of comparison preferred. The important thing is to give reasons for decisions.

The two questions (about the picture and about the dot plots) can be given to pairs or small groups to consider before a general discussion.

There may not always be a clear-cut answer.

◊ Pupils may suggest finding the mean. You could offer data where the mean of one group is greater than the mean of another, but only because of one extreme value (for example, 190, 150, 150, 150 and 159, 159, 159, 159). Work on the mean and discussion of which 'average' is more appropriate is dealt with later in the course.

37 Comparisons • 199

B **Median** (p 266)

◊ You can introduce the median using the pupils' own heights. They will need to know their heights or will need to measure them.

Start by getting an odd number of pupils to stand in order of height. Emphasise that it is not the middle person who is the median of the group, but that person's height. It is a good idea to use 'median' only as an adjective at first, for example 'median height', 'median age'.

Then do the same with an even number of pupils.

To emphasise the value of the median as a way of comparing data, you may wish to carry out this activity with two separate groups (boys and girls or sides of the class).

C **Range** (p 268)

◊ The range can be shown practically with a group of pupils. Ask the group to stand in height order. Ask the tallest and shortest to stand side by side. Measure the difference between their heights.

D **How fast do you react?** (p 270)

Pupils work in pairs.

| Sheet 141 |

◊ In addition to comparing performance within each pair, pupils could compare left and right hands. For homework they could compare themselves with an adult.

◊ In one class pupils felt that they were getting clues from twitching fingers just before the ruler was dropped, so a card was used to cover the fingers.

◊ You may wish to work through section F (on report writing) first and get pupils to write up their work on reaction times. If so, you will need to explain the diagrams in section E as well.

E **Summarising data** (p 271)

The diagram in the pupil's book is a simplified version of the 'box and whisker' diagram, which shows the quartiles as well.

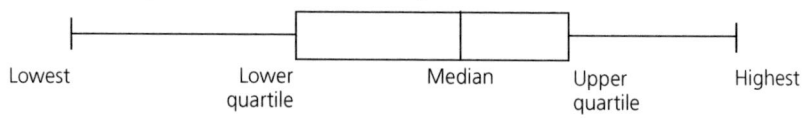

Lowest — Lower quartile — Median — Upper quartile — Highest

Either type of diagram may be arranged vertically.

Make sure pupils align the diagrams for each question correctly on a common scale, so that they can be used to make comparisons.

F Writing a report (p 272)

T

The specimen report is intended to help pupils write their own reports. It can be used for either group or class discussion

◊ You may need to make it clear that the fastest person is the one with the shortest time, and vice versa.

G Projects (p 274)

Each project generates data for making comparisons in a short written report.

The Argon Factor

Sheets 142 and 143 (preferably on OHP transparencies), sheet 144

T

◊ There are two tests: mental agility and memory.

◊ In the mental agility test, give the pupils one minute to memorise the shapes and numbers on sheet 142. (They are best shown on an OHP.)

Tell pupils 'You will be given 5 seconds to answer each question. Questions will be read twice'.

1. What number is inside the circle?
2. What number is inside the pentagon?
3. What number is inside the first shape?
4. What shape has the number 17 in it?
5. What shape has the number 29 in it?
6. What number is inside the middle shape?
7. What number is inside the last shape?
8. What shape is in the middle?
9. What is the shape before the end one?
10. What number is in the second shape?
11. What shape is the one after the one with 21?
12. What number is inside the fourth shape?
13. What number is in the shape just right of the triangle?
14. What shape is two to the left of the kite?
15. What number is in the shape two after the circle?

'Argon factor is excellent as a class activity and well worth the time spent on it.'

◊ In the memory test, give the pupils two minutes to remember the pictures and details of the four people on sheet 143. (You might want to use fewer pictures or less information with some classes.)

Then give them ten minutes to write answers to the 20 questions on sheet 144.

37 Comparisons • 201

◊ Discuss with the class what comparisons can be made from the scores. Suggestions include these.
- Do people remember more about their own sex?
- Which test did the class do better on? (Remember that the number of questions is not the same.)

Other projects

> For 'Tile pattern' (see below), sheet 145

Another project idea is given on p 274 ('Handwriting size')

Another possibility is 'Tile pattern', for which sheet 145 is needed.
This is suitable for an individual pupil or a pair. The pupil(s) carrying out the project cut out the 16 tiles and put them in an envelope. They decide which groups of people are to be compared (for example, children and adults). Each 'subject' is then timed making the pattern shown on the sheet.

B Median (p 266)

B1 (a) 159 cm (b) 156 cm (c) 154.5 cm

B2 (a) 11 (b) 154 cm

B3 A Girls 136 cm; boys 153 cm
The boys are taller.

B Girls 149.5 cm; boys 143 cm
The girls are generally taller.

C Girls 143 cm; boys 149 cm
The boys are generally taller but the girls' heights are well spread out.

D Girls 149 cm; boys 140 cm
The girls are generally taller but there are some short girls and one tall boy.

E Girls 139 cm; boys 149 cm
The boys are generally taller but there is a tall girl and some short boys.

F Girls 152 cm; boys 145 cm
The girls are generally taller.

B4 (a) 152 cm (b) 151 cm
(c) 153 cm (d) 152.5 cm

B5 (a) 69 kg (b) No change
(c) No change (d) Up by 1 kg

B6 (a) 52, 54, 58, 60, 63, 65, 70
(b) 60 kg

B7 (a) 152 cm (lengths in order are 139, 148, 152, 156, 161 cm)
(b) 36 kg (weights in order are 26, 29, 31, 34, 38, 39, 40, 45 kg)

B8 Boys have the greater median weight (2.7 kg); the girls' median is 2.65 kg.

C Range (p 268)

C1 (a) A 13 cm B 11 cm C 18 cm
(b) C (c) B

C2 (a) 13 minutes (b) 4 minutes
(c) 9 minutes

C3 (a) Herd B
(b) Herd A, because it has the greatest range
(c) Herd C, because it has the smallest range

C4 (a) Median 28, range 12
(b) Median 85, range 29
(c) **Nicky** and **Carol** both had high scores, but **Nicky**'s scores were the more spread out of the two.
(d) Nicky's scores and **Martin**'s scores were both spread out, but **Nicky** had the higher scores of the two.

202 • 37 Comparisons

(e) **Paul** and **Martin** were both bad players because they had **low** median scores.

(f) Paul was a consistent player because the range of his scores was **low**.

C5 (a) Northern: median 12 m, range 7 m
Southern: median 14.5 m, range 11 m

(b) Southern trees are taller. Their heights are more spread out.

C6 (a) Machine A: median 500 g, range 26 g
Machine B: median 498 g, range 5 g
Machine C: median 502 g, range 6 g
Machine D: median 514 g, range 25 g

(b) Machine B   (c) Machine D

(d) Machine C

(e) (Machine A) Inconsistent: it both underfilled and overfilled packs.

(f) Machine C. It usually put enough in a pack to avoid complaint without being too generous to the customer.

E Summarising data (p 271)

E1 The pupil's diagrams

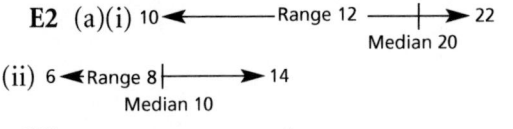

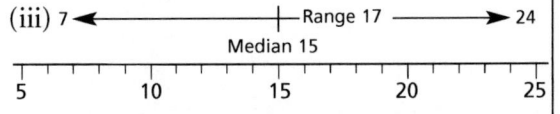

E2 (a)(i) 10 ◄──── Range 12 ────┼──► 22
              Median 20

(ii) 6 ◄Range 8┼────► 14
         Median 10

(iii) 7 ◄─────┼── Range 17 ────► 24
              Median 15

(b) Linford is consistently quick. He has the lowest median time and the smallest range.
Jules has some quick times, but is the least consistent, as shown by the large range.

What progress have you made? (p 274)

1 Left hand: median 18, range 16
Right hand: median 14, range 6
Pat is faster and less variable using his right hand.

Practice booklet

Sections A, B and C (p 84)

1 (a) 138 g   (b) 140 g   (c) 207 g

2 (a)

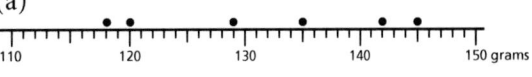

110 120 130 140 150 grams

(b) 132 g

3 (a) Barcelona 21°C Birmingham 22°C

(b) Birmingham

(c) Barcelona 12 degrees
Birmingham 8 degrees

Sections C and E (p 85)

1 The lengths are
A 27 mm   B 33 mm   C 36 mm
D 40 mm   E 56 mm   F 53 mm

Median length 38 mm
Range of lengths 26 mm

2 (a) Median 85 kg, range 46 kg
(b) Median 7.8 m, range 3.3 m
(c) Median 1°C, range 10°C
(d) Median 37.7°C, range 2.8°C

3 The pupil's comparison, such as 'The grey squirrels weigh more, because the median weight of the red squirrels is 293 g, and the median of the greys is 599 g. The greys are also more varied in weight, as the range of the reds is 25 g, but the range of the greys is 112 g.'

4 On the whole, Jo is faster (the medians are 80.1 s and 81.7 s). Jo is also more consistent (the ranges are 3.4 s and 8.3 s). But Jay has had the fastest single run (76.2 s).

5 Jo:
78.3 ◄─Range 3.4─┼────► 81.7
         Median 80.1
Jay:
76.2 ◄────Range 8.3─┼──► 84.5
         Median 81.7

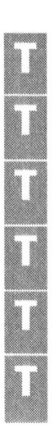

38 Multiplying and dividing

7T/15

The unit is to be done without a calculator.

p 275 **A** Multiplying by multiples of ten

p 276 **B** Table method

p 278 **C** Lining up columns

p 278 **D** Lattice method

p 279 **E** Division with no remainders

p 280 **F** Division with remainders

Practice booklet pages 86 and 87

A Multiplying by multiples of ten (p 275)

Pupils should see that multiplying by, say, 30 is equivalent to multiplying by 3 and then by 10.

B Table method (p 276)

Many pupils prefer the table method because they can see clearly where each number comes from. The standard 'long multiplication' method is often carried out wrongly when pupils lose sight of the reasons for the steps.

◊ You may wish to start with a simpler example such as 12 × 14.

For 23 × 34 some pupils may find this helpful:

×	10	10	10	4
10	100	100	100	40
10	100	100	100	40
3	30	30	30	12

C Lining up columns (p 278)

Pupils can try to relate the table method to the 'long multiplication' method.

204 • 38 Multiplying and dividing

D Lattice method (p 278)

This is of historical interest and some pupils may well prefer it to other methods.

◊ Another method which may be of interest is the so-called 'Egyptian' method, which is based on doubling. You can write the table below on the board and ask pupils if they can use the numbers to work out 47 × 59.

1	59
2	118
4	236
8	472
16	944
32	1888

By breaking 47 down into 32 + 8 + 4 + 2 + 1, they can work out 47 × 59 by adding 1888, 472, 236, 118 and 59.

Four digits

◊ Pupils may comment that
- with four different digits there are 12 different multiplications (counting 34 × 56 the same as 56 × 34):

  34 × 56   43 × 56   34 × 65   43 × 65   36 × 54   63 × 54
  36 × 45   63 × 45   46 × 53   64 × 53   46 × 35   64 × 35

- the largest result is 63 × 54 and the smallest is 46 × 35

◊ For the largest result, the positions of the 6 and the 5 in the tens place are obvious. The choice is then between 64 × 53 and 63 × 54. In the second case the larger units figure multiplies the larger tens figure, so this gives the larger result.

Five digits

◊ The largest result is with 752 × 84. The positions of the 8, 7, 5 and 4 follow from the four digits case. The 2 could either make 842 × 75 or 84 × 752. In the second case 2 is multiplied by 84 and in the first by 75, so the second is better.

◊ The smallest result is with 478 × 25.

E Division with no remainders (p 279)

◊ In the introductory discussion make sure that pupils appreciate that 368 ÷ 23 can be thought of as '368 shared between 23' (sharing) as well as 'how many 23s in 368' (grouping).

◊ Pupils may have learned the standard 'long division' algorithm and be happy with it. But many do not find it straightforward and the 'chunking' method (Sheena's method) is a fairly efficient alternative.

◊ It may be worth practising mentally working out 13 × 10, 13 × 20, 13 × 5 as this is useful when using the 'chunking' method.

◊ The example 325 ÷ 13 can be used to illustrate dealing first with a chunk of twenty 13s, followed by a chunk of five 13s.

F Division with remainders (p 280)

◊ In the initial discussion pupils should understand that in the example, when the final chunk of 52 is taken from 91, 39 is left over and is the remainder.

A Multiplying by multiples of ten (p 275)

A1 (a) 80 (b) 120 (c) 300
     (d) 2100 (e) 2000

A2 The pupil's explanation of how to convince someone that 40 × 30 ≠ 120

A3 (a) 800 (b) 900 (c) 1200
     (d) 1500 (e) 1400 (f) 2000
     (g) 1000 (h) 6300 (i) 4800
     (j) 4000

A4 (a) 6000 (b) 12 000 (c) 8000
     (d) 42 000 (e) 2700 (f) 2000
     (g) 48 000 (h) 20 000 (i) 60 000
     (j) 30 000

A5 (a) 60 (b) 50 (c) 900
     (d) 8 (e) 400 (f) 5

A6 20 × 20 = 400
     20 × 40 = 800
     20 × 80 = 1600
     20 × 160 = 3200
     20 × 400 = 8000
     40 × 40 = 1600
     40 × 80 = 3200

A7 (a) 620 (b) 4500 (c) 5400
     (d) 1440 (e) 85 600

B Table method (p 276)

B1 13 × 12

×	10	2		100
10	100	20	+	20
3	30	6	+	30
			+	6
				156

B2 14 × 15

×	10	5		100
10	100	50	+	50
4	40	20	+	40
			+	20
				210

B3 16 × 23

×	20	3		200
10	200	30	+	30
6	120	18	+	120
			+	18
				368

B4 24 × 35

×	30	5		600
20	600	100	+	100
4	120	20	+	120
			+	20
				840

B5 (a) (i) C (40 × 30)
         (ii) 1200
     (b) 1204

B6 (a) 1800 (b) 1767

B7 (a) 585 (b) 1512 (c) 1692
     (d) 2916 (e) 2475

206 • 38 Multiplying and dividing

B8 53 × 148

×	100	40	8
50	5000	2000	400
3	300	120	24

```
  5000
+ 2000
+  400
+  300
+  120
+   24
  ————
  7844
```

B9 362 × 47

×	40	7
300	12000	2100
60	2400	420
2	80	14

```
 12000
+ 2100
+ 2400
+  420
+   80
+   14
 —————
 17014
```

B10 (a) (i) D (700 × 40)
     (ii) 28 000
  (b) 25 197

B11 (a) 2950   (b) 3936   (c) 3103
    (d) 13 496 (e) 40 420

C Lining up columns (p 278)

C1 (a) 585   (b) 1512   (c) 1692
    (d) 2756 (e) 2475 (f) 2916
    (g) 3564 (h) 5166

C2 (a) The pupil's explanation such as 'The pupil has multiplied 16 by 2 rather than 20.'
    (b) 384

D Lattice method (p 278)

D1 (a) 13 × 14 = 182

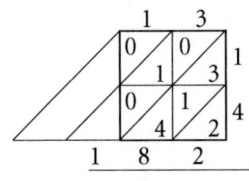

(b) 24 × 38 = 912

(c) 67 × 42 = 2814

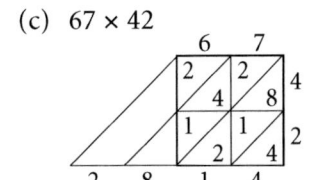

(d) 215 × 34 = 7310

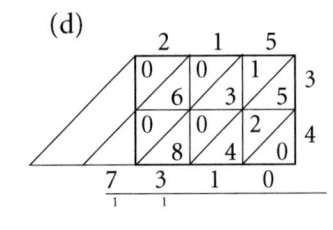

(e) 35 × 107 = 3745

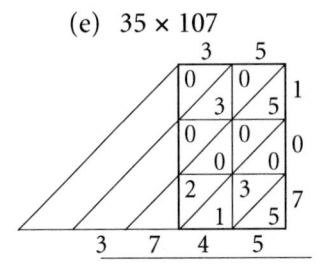

(f) 246 × 458 = 112 668

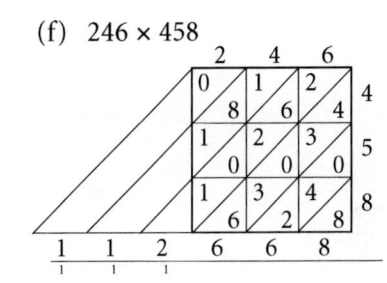

D2 (a) 63 × 37 = 2331
    (b) 37 × 63 = 2331

E Division with no remainders (p 279)

E1 (a) 6 (b) 8 (c) 7 (d) 8
    (e) 12 (f) 16 (g) 18 (h) 21

E2 25

E3 21

E4 28

E5 £7

E6 £56

E7 76

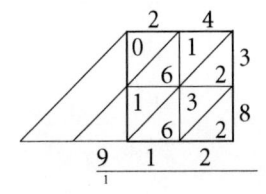

E8 The pupil's problem in words for 594 ÷ 18

E9 (a)–(c) The pupil's problems in words

E10 (a) 21 (b) 17 (c) 38

F Division with remainders (p 280)

F1 (a) 12 r 3 (b) 11 r 19 (c) 19 r 5
(d) 16 r 23 (e) 19 r 16 (f) 9 r 12
(g) 15 r 24 (h) 26 r 46

F2 CABBAGE

F3 The pupil's word puzzle

F4 Two of them do (21 and 23).

F5 He needs 17 packets.
He will have 6 more rolls than he needs.

F6 He can make 21 necklaces.
He will have 34 beads left over.

F7 1279

F8 48

F9 1

F10 155 seconds

F11 39 minutes and 10 seconds

F12 4 hours and 40 minutes

What progress have you made? (p 282)

1 (a) 1200 (b) 30 000 (c) 24 000
(d) 120 000 (e) 4800 (f) 1520

2 (a) 2000 (b) 1887

3 (a) 234 (b) 399 (c) 1938
(d) 1288 (e) 10 810 (f) 7448

4 £43.20

5 (a) 22 (b) 13 (c) 19

6 £36

7 (a) 17 r 11 (b) 23 r 10 (c) 41 r 1

8 28

Practice booklet

Sections A and B (p 86)

1 (a) 1400 (b) 12 000
(c) 240 000 (d) 4000

2 (a) 192

×	10	6
10	**100**	**60**
2	**20**	**12**

(b) 182

×	10	4
10	**100**	**40**
3	**30**	**12**

(c) 408

×	20	4
10	**200**	**40**
7	**140**	**28**

(d) 2170

×	60	2
30	**1800**	**60**
5	**300**	**10**

(e) 12 792

×	200	40	6
50	**10 000**	**2000**	**300**
2	**400**	**80**	**12**

(f) 21 105

×	300	30	5
60	**18 000**	**1800**	**300**
3	**900**	**90**	**15**

3 The pupil's tables leading to these answers
 (a) 255 (b) 1665
 (c) 7344 (d) 10 224

Sections C and D (p 86)

1 774

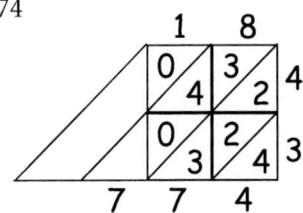

```
    1   8
   0 3
  4 2 4
   0 2
  3 4 3
  7 7 4
```

2 (a) 378 (b) 1568
 (c) 2209 (d) 1350

3 The largest is 81 × 64 = 5184.
 The smallest is 16 × 48 = 768.

4 (a) 6164 (b) 10 664
 (c) 97 812 (d) 104 755

Sections E and F (p 87)

1 (a) 4 (b) 11 (c) 15 (d) 13
 (e) 25 (f) 28 (g) 32 (h) 21

2 12

3 £26

4 (a) 6 r 11 (b) 12 r 2
 (c) 21 r 10 (d) 19 r 21

5 12 r 3 (C), 17 r 1 (A), 24 r 8 (H), 25 r 5 (E), 11 r 2 (B)
 BEACH

6 14 packets are needed.
 There will be 16 spare pencils

7 They can make 14 chains.
 There will be 15 links left over.

38 Multiplying and dividing • 209

39 Think of a number

7C/32

This unit develops the important idea of an inverse process and shows how to use this idea to solve an equation. 'Think of a number' puzzles are the context.

In later work, pupils should realise the limitations of using arrow diagrams to solve equations. The 'balancing' method should then become the principal method for solving equations.

In this unit, arrow diagrams are drawn with circles and ellipses. Pupils may find it easier to use squares and rectangles in their diagrams.

p 283 **A** Number puzzles	Using arrow diagrams and inverses to solve 'think of a number' puzzles
p 286 **B** Using letters	Writing number puzzles as equations
p 288 **C** Solving equations	Solving linear equations using arrows diagrams and inverses
p 289 **D** Quick solve	A game to consolidate solving equations

Essential	**Optional**
Calculators	Sheet 160
Sheet 159	
Practice booklet pages 88 to 90	

A Number puzzles (p 283)

Calculators for working with decimals and large numbers

◊ You could introduce this section by asking the pupils to each think of a number without telling anyone what it is.

Now ask them to
 Add 1.
 Multiply by 3.
 Add 5.
 Take away 2.
 Divide by 2.

Ask some pupils to tell you what their answers are and then work backwards to give the numbers they were thinking of.

'This introduction went down very well. By giving simple numbers and carefully timing when I asked them I was able to get even the weakest to respond.'

Pupils could discuss how they think you worked them out.

Now, give some single-operation problems such as

> I think of a number.
> I multiply by 9 and my answer is 216.
> What number did I think of?

Ensure discussion of these brings out the idea of using an inverse and arrow diagrams.

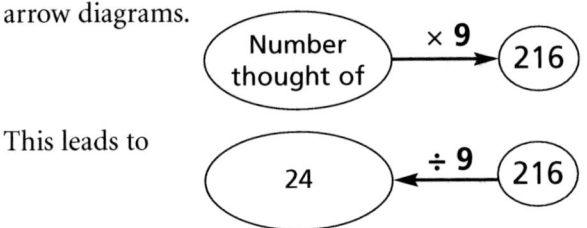

$$\text{Number thought of} \xrightarrow{\times 9} 216$$

This leads to

$$24 \xleftarrow{\div 9} 216$$

Encourage pupils to check each solution by substituting in the original puzzle.

Now move on to 'think of a number' problems that use more than one operation and show how arrow diagrams and inverses can be used to solve them.

Many pupils will feel more confident using 'trial and improvement' to solve these problems. To demonstrate the power of using inverses, include problems involving decimals and many operations. For example

> I think of a number.
> I divide by 5.
> I subtract 45.
> I multiply by 1.2.
> I add 8.
> I multiply by 0.2.
> I subtract 12.
> My answer is 10.
> What number did I think of?

Pupils could use both trial-and-improvement and inverse methods and compare them.

You could end this teacher-led session by asking the pupils to solve the puzzles on the pupil page. The solutions to these puzzles are 17, 12 and 3.3 respectively.

B **Using letters** (p 286)

'Very good ... needed to refresh use of brackets first.'

It may help to tell pupils that this section is not about solving the puzzles or equations, but about linking equations, arrow diagrams and puzzles. Otherwise, they may feel that they have not fully answered the questions unless they have 'solved' each puzzle to find the number thought of.

The questions could be used as a basis for a whole-class discussion.

◊ Include some examples where multiplication is the first operation and some where it is the second operation. Pupils may take some time to understand when they need to use brackets. You could include the puzzle and diagrams leading to the equation $2n + 3 = 20$ bringing out how it differs from $2(n + 3) = 20$ (the example on page 286).

◊ Make sure pupils know that $2 \times (n + 3)$ is the same as $(n + 3) \times 2$ and that $2(n + 3)$ is shorthand for it.

B1 Watch out for pupils who match up A with X to start with.

C Solving equations (p 288)

Calculators for working with decimals and large numbers

◊ In your discussion, include the equation $2n - 3 = 130$ and compare it with $2(n - 3) = 130$ on page 288.

◊ The equations in question C5 involve more than two operations. It may be beneficial to include some examples of this kind in your introduction.

D Quick solve (p 289)

The version of this game described in the pupil's book can be played in groups of three or four. It can also be played as a whole class or individually (see below).

Each group needs a set of 36 cards (sheet 159).
Optional: Each pupil needs a copy of sheet 160 if they check each other's answers. Note that sheets 159 and 160 look similar to sheets 157 and 158. Care should be taken not to confuse them.

◊ Emphasise that *all* players take a card at the start of the game and take another as soon as they think they have solved their equation. They should keep their equation cards – they will not be used by another player.

◊ There are other ways to use the cards.

One pile version

This is played as the version in the book but pupils take cards from a mixed shuffled pile.

Three pile whole-class version

The game could be played as a whole class with the teacher having sets of cards in three piles: cards worth 1 point, 2 points and 3 points.

212 • 39 *Think of a number*

Individual pupils ask the teacher for a 1, 2 or 3 point card. When they think they have solved the equation, they ask for another card.

Continue until all the cards have been taken or some specified time limit has been reached.

One pile whole-class version

This is played as the three pile whole-class version but pupils take cards at random from a mixed pile.

Individual version

The cards do not need to be cut out for this version. Each pupil solves as many equations as they can from the set of 36 in a specified time. Solutions are checked and points awarded as before.

The game can be played with the number of points awarded for a correct solution being the value of the solution.

A set of solutions is given below.

Card 1	$n = 2$	Card 19	$n = 13.5$
Card 2	$n = 5$	Card 20	$n = 30$
Card 3	$n = 4$	Card 21	$n = 18$
Card 4	$n = 18$	Card 22	$n = 25$
Card 5	$n = 3$	Card 23	$n = 0$
Card 6	$n = 4$	Card 24	$n = 27$
Card 7	$n = 5$	Card 25	$n = 7.6$
Card 8	$n = 4$	Card 26	$n = 3.8$
Card 9	$n = 3$	Card 27	$n = 12$
Card 10	$n = 1$	Card 28	$n = 3.2$
Card 11	$n = 6$	Card 29	$n = 2$
Card 12	$n = 3$	Card 30	$n = 6.5$
Card 13	$n = 1.5$	Card 31	$n = 10$
Card 14	$n = 2.5$	Card 32	$n = 3$
Card 15	$n = 10$	Card 33	$n = 3$
Card 16	$n = 13$	Card 34	$n = 6.4$
Card 17	$n = 9$	Card 35	$n = 2.3$
Card 18	$n = 0.5$	Card 36	$n = 2.5$

A Number puzzles (p 283)

A1 17 ←+4— 13 ←÷3— 39

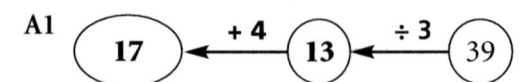

The number thought of was 17.

A2 (a) Puzzle 1: C   Puzzle 2: B

(b) Puzzle 1

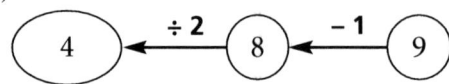

4 ←÷2— 8 ←−1— 9

The number thought of was 4.

Puzzle 2

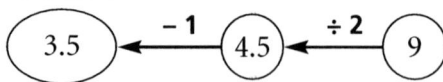

3.5 ←−1— 4.5 ←÷2— 9

The number thought of was 3.5.

A3 (a) I think of a number.
- I divide by 5.
- I take away 7.

The result is 17.

What number did I think of?

(b) 120 ←×5— 24 ←+7— 17

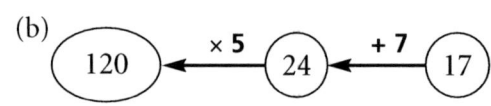

The number thought of was 120.

A4 (a)

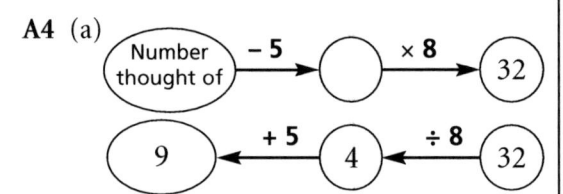

Number thought of —−5→ ○ —×8→ 32

9 ←+5— 4 ←÷8— 32

The number thought of was 9.

(b)

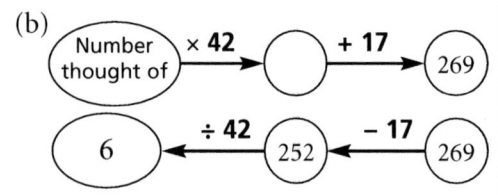

Number thought of —×42→ ○ —+17→ 269

6 ←÷42— 252 ←−17— 269

The number thought of was 6.

(c)

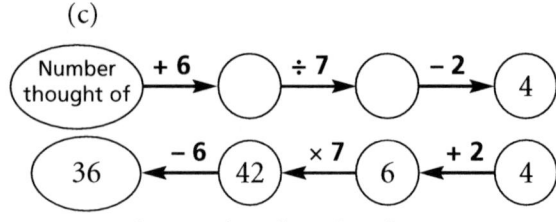

Number thought of —+6→ ○ —÷7→ ○ —−2→ 4

36 ←−6— 42 ←×7— 6 ←+2— 4

The number thought of was 36.

(d)

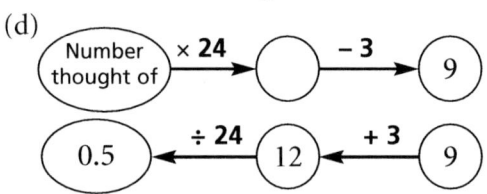

Number thought of —×24→ ○ —−3→ 9

0.5 ←÷24— 12 ←+3— 9

The number thought of was 0.5.

(e)

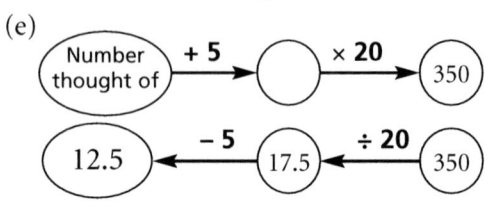

Number thought of —+5→ ○ —×20→ 350

12.5 ←−5— 17.5 ←÷20— 350

The number thought of was 12.5.

A5 Using positive whole numbers and zero gives the following possibilities.
- I add 10, I multiply by 1.
- I add 4, I multiply by 2.
- I add 2, I multiply by 3.
- I add 1, I multiply by 4.
- I add 0, I multiply by 6.

Extending to negative numbers and decimals gives an infinite number of possibilities.

A6 The pupil's puzzles

B Using letters (p 286)

B1 (a) A: Z, B: V, C: Y, D: X

(b) n —+1→ ○ —×2→ 15

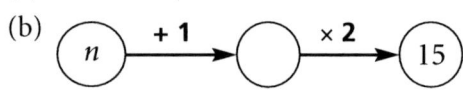

B2 (a) Puzzle 1: E   Puzzle 2: C
Puzzle 3: B   Puzzle 4: D

(b) I think of a number.
- I subtract 2.
- I multiply by 5.

The result is 8.

What number did I think of?

B3 (a) $3n + 4 = 108$
   (b) $16n = 52$
   (c) $6(n - 5) = 42$

B4 (a) $5(n + 3) = 20$
   (b) $9k - 3 = 15$

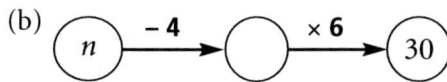

B5 (a) $n \xrightarrow{\times 3} \bigcirc \xrightarrow{+ 1} 22$

   (b) $n \xrightarrow{- 4} \bigcirc \xrightarrow{\times 6} 30$

C Solving equations (p 288)

C1 (a) $n = 14$ (b) $h = 25$ (c) $m = 21$
   (d) $x = 13$ (e) $p = 16$ (f) $z = 5$

C2 (a) $m = 2.5$ (b) $a = 3.5$ (c) $n = 4.2$
   (d) $p = 5.6$ (e) $x = 7.5$ (f) $y = 12.6$

C3 $(2 \times 6) + 4 = 12 + 4 = 16$
   The pupil's equations with solution $n = 6$

C4 The pupil's equations with solution $y = 1.5$

*C5 (a) $m = 23$ (b) $q = 5$ (c) $p = 0.05$

What progress have you made? (p 290)

1 (a) Number thought of $\xrightarrow{\times 3} \bigcirc \xrightarrow{- 5} 82$
   $29 \xleftarrow{\div 3} 87 \xleftarrow{+ 5} 82$

   The number thought of was 29.

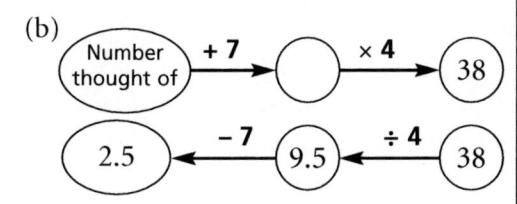

   (b) Number thought of $\xrightarrow{+ 7} \bigcirc \xrightarrow{\times 4} 38$
   $2.5 \xleftarrow{- 7} 9.5 \xleftarrow{\div 4} 38$

   The number thought of was 2.5.

2 A: Z B: W C: Y

3 (a) $z = 37$ (b) $p = 32$ (c) $y = 5.5$
   (d) $x = 1.2$ (e) $q = 85$

Practice booklet

Section A (p 88)

The pupil's arrow diagrams leading to the following answers:

1 11   2 31
3 7   4 136
5 2.5   6 0.5
7 21   8 1.5

Section B (p 89)

1 (a) $14n = 56$ (b) $5n - 7 = 103$
  (c) $3(n + 4) = 54$ (d) $17(n - 8) = 68$

2 (a) $7p - 4 = 31$ (b) $3(s + 8) = 33$
  (c) $6(q - 5) = 60$ (d) $4m - 11 = 37$

*3 $4(11n + 2) = 96$

Section C (p 90)

1 The pupil's arrow diagrams leading to the following answers:
   (a) $x = 9$ (b) $h = 12$ (c) $d = 7$
   (d) $k = 19$ (e) $p = 19$ (f) $s = 7$
   (g) $q = 4$ (h) $m = 11$ (i) $v = 21$
   (j) $n = 15$

2 (a) $y = 2.5$ (b) $w = 4.5$ (c) $h = 6.5$
   (d) $b = 9.5$ (e) $c = 1.5$ (f) $z = 3.6$
   (g) $p = 1.8$ (h) $q = 3.5$ (i) $t = 1.3$
   (j) $x = 2.7$

3 (a) $n = 5.1$ (b) $a = 3.3$ (c) $d = 2.4$
   (d) $c = 8.2$ (e) $a = 1.5$ (f) $w = 6.5$
   (g) $h = 6.1$ (h) $p = 0.2$ (i) $q = 6.9$
   (j) $x = 2.5$

4 The pupil's equations with solution $m = 5$

5 The pupil's equations with solution $x = 1$

6 The pupil's equations with solution $p = 2.5$

40 Symmetry 2 7C/2

p 291 **A** What is symmetrical about these shapes?	Discussing reflection and rotation symmetry
p 292 **B** Rotation symmetry	Identifying rotation symmetry Centres and orders of rotation symmetry
p 293 **C** Making designs	Making patterns with rotation symmetry
p 295 **D** Rotation and reflection symmetry	Identifying rotation and reflection symmetry
p 297 **E** Pentominoes	Making shapes with reflection and/or rotation symmetry

Essential

Tracing paper
Mirrors
Square dotty paper
Triangular dotty paper
Sheets 117, 120, 121 and 122

Practice booklet pages 91 to 94

Optional

OHP transparency made from sheet 116

A What is symmetrical about these shapes? (p 291)

Discussion should consolidate earlier work on reflection symmetry and show you how much they know already about rotation symmetry.

Optional: OHP transparency made from sheet 116, tracing paper

◊ One way to generate discussion is for pupils to study the page individually, then discuss it in small groups; then you can bring the whole class together and ask for contributions from the groups.

◊ Pupils should be able to describe the reflection symmetry of the shapes. Some may realise that shapes with only rotation symmetry (B, C, D, I) are symmetrical in some way but be unable to describe how. Others may know about rotation symmetry already.

216 • 40 Symmetry 2

B **Rotation symmetry** (p 292)

> Tracing paper, sheet 117
> Optional: Square dotty paper

◊ The shape on the page is the first one on sheet 117.

◊ A tracing can be rotated by putting a pencil point at the centre of rotation.

C **Making designs** (p 293)

> Tracing paper, square dotty paper, triangular dotty paper, sheets 120 and 121

C1 You may need to go through the instructions. Check that pupils are rotating the tracing paper to get the new position of the shape. Some may be flipping it over.

◊ Pupils could try the subsequent activities without tracing paper and use tracing paper to check.

D **Rotation and reflection symmetry** (p 295)

> Tracing paper, mirrors, sheet 122, square dotty paper

E **Pentominoes** (p 297)

> Tracing paper, mirrors, square dotty paper

B Rotation symmetry (p 292)

B1 Centres of rotation marked on sheet 117

Shapes with rotation symmetry	Order of rotation symmetry
A	4
B	3
C	2
E	4
F	2
G	3
I	2
J	6
K	8
L	2
M	4

C Making designs (p 293)

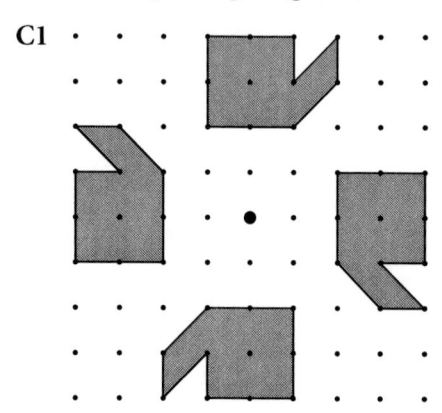

C1

C2 Completed designs on sheet 120

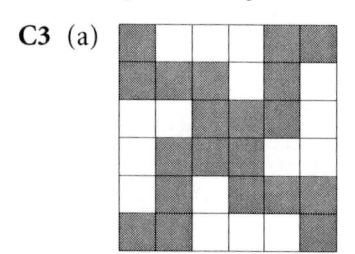

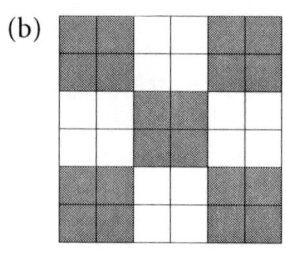

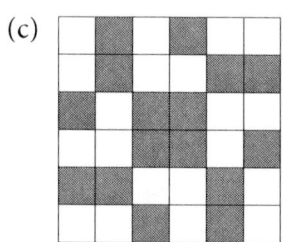

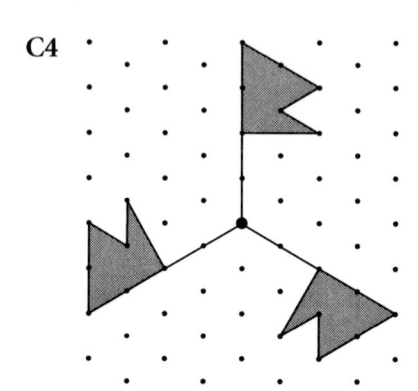

C3 (a)

(b)

(c)

C4

C5 Completed designs on sheet 121

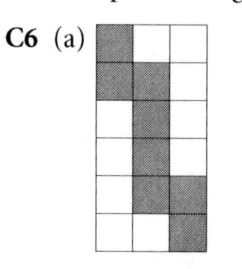

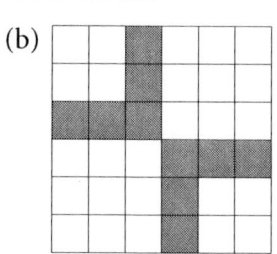

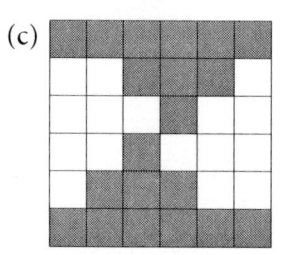

C6 (a) (b)

(c)

*****C7** The pupil's design with rotation symmetry of order 6

D Rotation and reflection symmetry (p 295)

D1

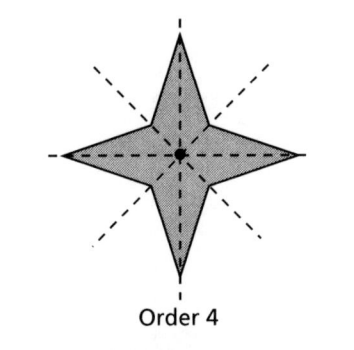

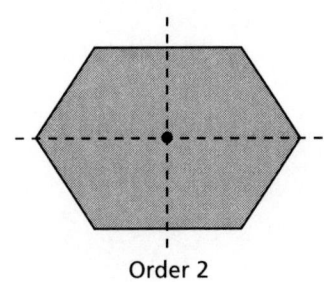

Order 4

Order 2

D2 (a) A Lines 1 and 3
  B Lines 1, 2, 3 and 4 (all of them)
  C Neither is a line of symmetry
  D Lines 1, 3 and 5

(b) A Order 2
  B Order 4
  C Order 2
  D Order 3

D3 Centres of rotation and lines of symmetry marked on sheet 122

(a) Rotation symmetry of order 2
  No lines of symmetry

(b) Rotation symmetry of order 3
  Three lines of symmetry

(c) No rotation symmetry
  No lines of symmetry

(d) No rotation symmetry
  One line of symmetry

(e) Rotation symmetry of order 3
  No lines of symmetry

(f) Rotation symmetry of order 2
  Two lines of symmetry

(g) Rotation symmetry of order 2
  No lines of symmetry

(h) Rotation symmetry of order 2
  Two lines of symmetry

(i) No rotation symmetry
  One line of symmetry

(j) No rotation symmetry
  No lines of symmetry

(k) Rotation symmetry of order 2
  No lines of symmetry

(l) No rotation symmetry
  No lines of symmetry

(m) No rotation symmetry
  No lines of symmetry

(n) Rotation symmetry of order 2
  No lines of symmetry

D4 (a) 10:01:18, 18:08:03

(b) 11:11:11, 18:11:81

(c) The pupil's date with two lines of symmetry

(d) 16:11:91, 11:11:11, 18:11:81

(e) The pupil's date with rotation symmetry

(f) (i) 04:02:33 – no reflection or rotation symmetry
  (ii) 31:10:81 – one line of symmetry but no rotation symmetry
  (iii) 08:11:80 – two lines of symmetry and rotation symmetry
  (iv) 09:11:60 – no lines of symmetry but rotation symmetry

D5 (a) Order 2

(b) Yes, two lines of symmetry

D6 There are sixteen ways to shade four squares to make a pattern with rotation symmetry (plus twelve that are 90° rotations of some of the sixteen). Pupils have to find eight different ways.

The sixteen ways are shown below.

Rotation symmetry of order 2

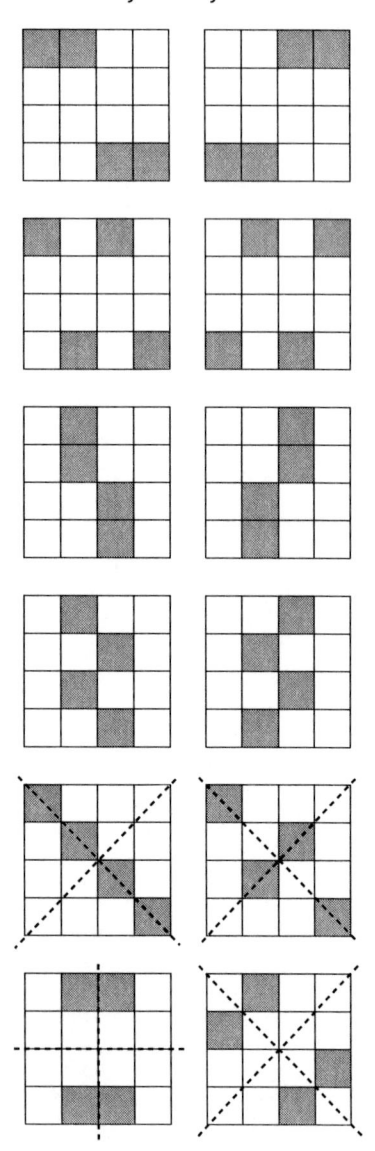

Rotation symmetry of order 4

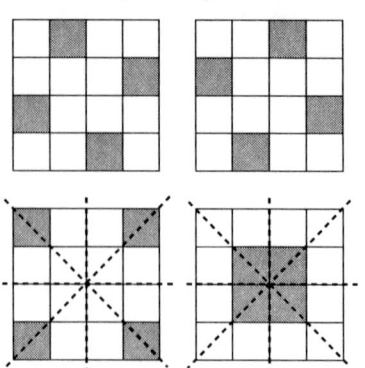

D7 (a) The pupil's pattern from

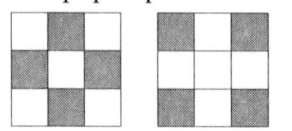

(b) The pupil's pattern from

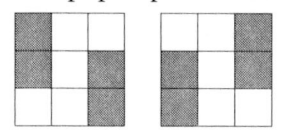

or 90° rotations of these

(c) Some examples are

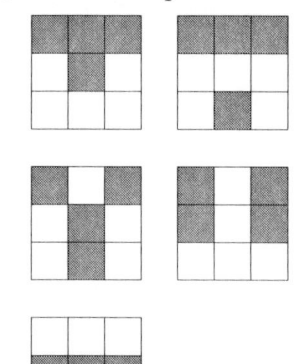

(d) Some examples are

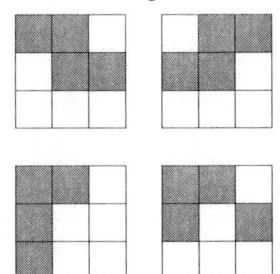

E Pentominoes (p 297)

E1 The pupil's pentomino from

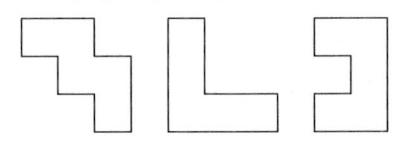

E2 (a) The pupil's pentomino from

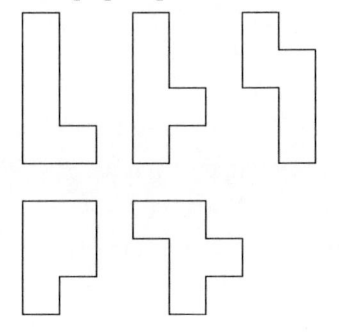

(b)   (c)

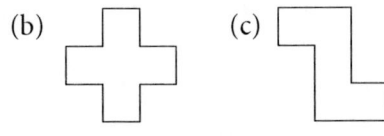

(d)

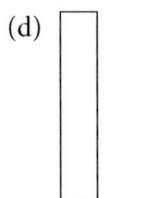

E3 (a) (i) The pupil's shape; examples are

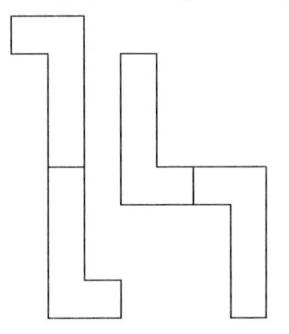

(ii) The pupil's shape; examples are

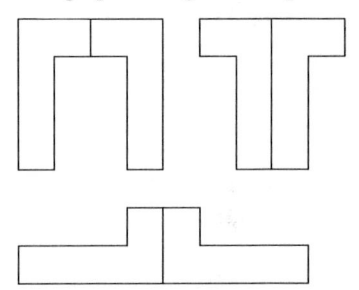

(iii) The pupil's shape; examples are

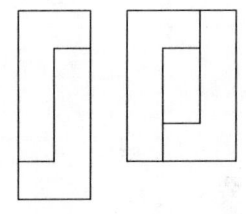

(b) The pupil's design; examples are

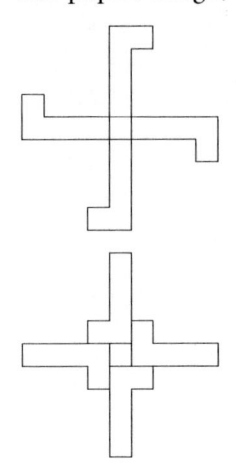

40 Symmetry 2 • 221

E4 The pupil's designs with
  (a) reflection symmetry but no rotation symmetry
  (b) rotation symmetry but no reflection symmetry
  (c) reflection symmetry and rotation symmetry

What progress have you made? (p 298)

1 (a) B and E
   (b) A
   (c) C and D

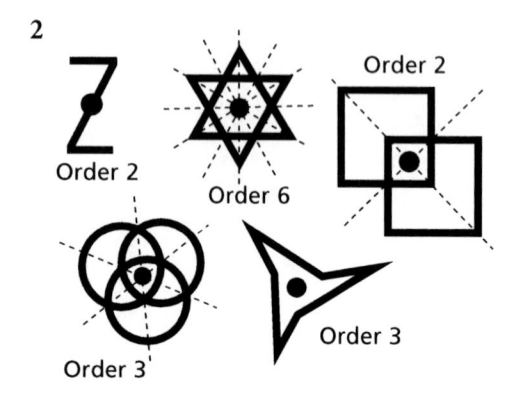

2 Z – Order 2
   Star – Order 6
   Overlapping squares – Order 2
   Three circles – Order 3
   Triangular shape – Order 3

3 The pupil's pattern; examples are

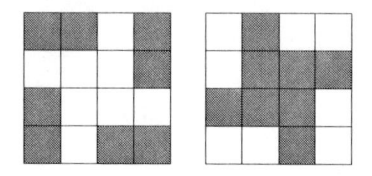

Practice booklet

Sections A and B (p 91)

1 Designs A, C and D have rotation symmetry.

2 (a) 5   (b) 8   (c) 2   (d) 4
   (e) 3   (f) 3   (g) 2   (h) 2

Section C (p 92)

1 The pupil's completed designs with rotation symmetry of order 4
2 The pupil's completed designs with rotation symmetry of order 2
3 The pupil's completed designs with rotation symmetry of order 3

Sections D and E (p 93)

1 (a) A 1, 2, 3, 4, 5
       B No lines of symmetry
       C 1, 5
   (b) A 5
       B 2
       C 2

2 (a) No lines of symmetry
       Rotation symmetry of order 3
   (b) Two lines of symmetry
       Rotation symmetry of order 2
   (c) One line of symmetry
       No rotation symmetry
   (d) No lines of symmetry
       No rotation symmetry

3 (a) No lines of symmetry
       Rotation symmetry of order 2
   (b) One line of symmetry
       No rotation symmetry
   (c) No lines of symmetry
       No rotation symmetry
   (d) No lines of symmetry
       Rotation symmetry of order 2

4 (a)

Rotation symmetry of order 4 and 4 lines of symmetry

No rotation symmetry or reflection symmetry

No rotation symmetry and 1 line of symmetry

Rotation symmetry of order 2 and 2 lines of symmetry

Rotation symmetry of order 2 and no reflection symmetry

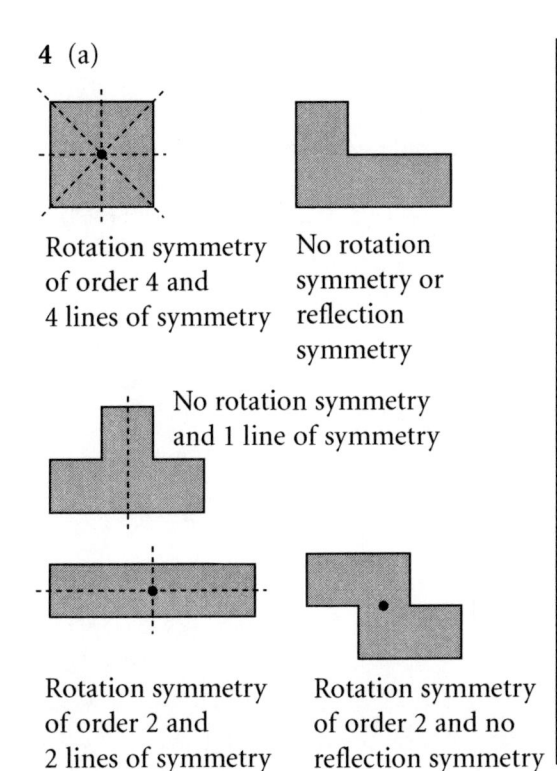

(b) It has one line of reflection symmetry and no rotation symmetry.

(c) Examples are
  (i)
  (ii)

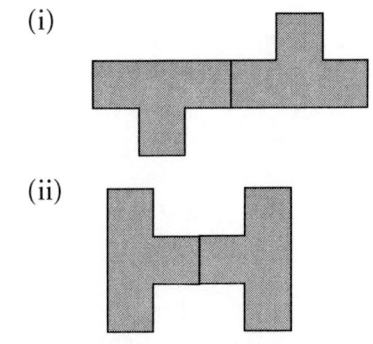

(d) The pupil's designs with
  (i) reflection symmetry but no rotation symmetry
  (ii) reflection symmetry and rotation symmetry
  (iii) rotation symmetry of order 4

㊸ Hot and cold (p 299)

This activity involves a lot of arithmetic and needs organisational skills.

Pupils imagine themselves to be running a stall selling ice creams and hot dogs. (In the simpler introductory version, there are only ice creams.) They get a forecast of tomorrow's temperature and have to decide how many ices and hot dogs to order. Then they find out the actual temperature, and how many of each they sell. (The forecast and actual temperatures are decided by the teacher throwing a dice.) Whatever isn't sold is wasted. The object, obviously, is to make as much profit as possible.

> Sheets 146 and 147 (one per pair)
> A dice

The activity is teacher-led, with pupils working in pairs, each pair running a stall.

◊ The simpler version, using sheet 146, is described here. The full version is similar, but with ice creams and hot dogs (sheet 147).

◊ Each pair starts with the same amount of money (for example £10). They record this in the Monday 'cash in box at start of day' cell.

◊ You throw the dice and announce the temperature forecast for Monday, as follows:

　　1 or 2: 10°C　　3 or 4: 15°C　　5 or 6: 20°C

◊ Each pair now has to decide how many ice creams (at 40p each) to order. Explain that the actual temperature could be above or below the forecast, as follows:

　　Forecast: 10°C　　Actual will be 7°C or 12°C
　　Forecast: 15°C　　Actual will be 12°C or 17°C
　　Forecast: 20°C　　Actual will be 17°C or 22°C

The number sold will depend on the actual temperature:

　　7°C　　0 sold　　　　12°C　　10 sold
　　17°C　 15 sold　　　 22°C　　30 sold

When pupils have decided how many ices to order, they fill in the order on the sheet and work out the cash left in the box.

◊ Now you throw the dice again. If it lands odd, you go down from the forecast to get the actual temperature (for example, 15°C forecast becomes 12°C actual); if even, you go up.

◊ Each pair works out its sales, at 90p each. (They can't sell more than they ordered!) They work out the cash in their box at the end of the day. Then you start another round.

T

> 'I tried this with top and bottom set. Both were hugely motivated and wanted to have several goes at it to improve their performances. We carried cash and supplies through week and worked out profits at end of week and had class winners'.

42 Investigations

7C/29

Investigative and problem-solving work are best integrated into the development of mathematical concepts and skills. However, focusing on investigative work, as here, allows important skills of report writing to be developed. It is not intended that all the investigations should be done together or in the order given.

Essential	**Optional**
Counters, tiles or pieces of paper (for B3)	Counters, tiles or pieces of paper (for B1) Square dotty paper (for B6)

A Crossing points (p 300)

Discussion of the report is intended to highlight some important processes, for example, specialising, tabulating, generalising, predicting, checking, explaining, drawing conclusions. It is not intended to suggest there is one 'correct' way to approach an investigation and write up the findings.

◊ If pupils find it difficult to follow Chris's written work, try to involve them actively. They could read through the first half of page 301 (the 1, 2 and 3 line results) and then try with 4 lines.
Emphasise that the lines should be drawn as long as possible so that all crossings are shown.

Ask pupils to find as many different numbers of crossings as possible with 4 lines; 0, 1, 3, 4, 5 and 6 are possible:

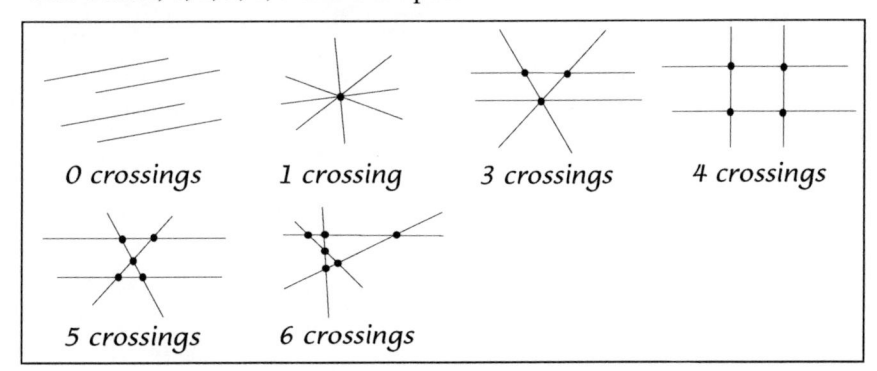

0 crossings   1 crossing   3 crossings   4 crossings

5 crossings   6 crossings

Ask how we can be sure that 2 crossings cannot be achieved or that 6 is the maximum number. Emphasise that pupils should try to explain their findings wherever possible.

◊ Now discuss how the investigation could proceed. Looking at the maximum number of crossing points is one choice (and it's the one made by the pupil in the write-up). Look at the results in the table. Pupils could try to spot a pattern before turning to page 303.

◊ Look at the table on page 303 and ask pupils if they can explain why the numbers of crossing points go up in the way they do. This is easier when pupils have found these results out for themselves. To get the maximum number of crossing points each additional line needs to cross all the lines in the diagram (except itself). Chris does not try to explain this in her report but without it she cannot be sure that the sequence of numbers continues in the way she describes. Point out that 'predict and check' may help to confirm results but is not foolproof (suppose there were actually 16 crossing points for 6 lines and your 'pattern' stopped you looking any further than 15).

◊ You could ask what the maximum number of crossing points would be with, say, 100 lines. Using Chris's method, this would take some time. With n lines, each line crosses $n-1$ lines to produce $n-1$ crossing points. However, $n(n-1)$ counts each crossing point twice so the number of crossing points is $\frac{n(n-1)}{2}$.

◊ As a further investigation, pupils can look at the maximum number of closed regions obtained. An interesting result is that you get the same set of numbers but with 0 included: 0, 1, 3, 6, 10, 15, …

B **Some ideas**

B1 Round table (p 304)

> Optional: Counters, tiles or pieces of paper may be useful to represent the people round the table.

◊ Pupils may find it helpful to do the investigation by moving counters or tiles (labelled A to E) round a drawing of a table. There is only one other arrangement:

```
      A
   C     D
      E  B
```

Ask pupils to consider how they know there are no other arrangements. They may realise that every person has sat next to every other person so no other arrangements are possible. Ask pupils if that is what they expected – often pupils think there will be more possibilities.

◊ In one school, the investigation proved easier when the table was 'unrolled' into a straight line (remembering that the two end people are in fact sitting next to each other). For five people the two different arrangements are

    ABCDE   ADBEC

◊ The results for 3 to 10 people are

Number of people	Numbers of arrangements
3	1
4	1
5	2
6	2
7	3
8	3
9	4
10	4

For an even number of people the rule is $a = \frac{p-2}{2}$
and for an odd number of people the rule is $a = \frac{p-1}{2}$
where a is the number of arrangements
and p is the number of people.

Since each person has two neighbours, the number of arrangements has to be the number of complete pairs of people that can sit next to an individual. These formulas give the number of complete pairs.

Very few pupils are likely to express the results algebraically at this stage.

B2 Nine lines (p 304)

◊ Clarify that the grids of squares have to be drawn with straight lines either parallel to each other or at right angles.

As in *Crossing points* lines should be drawn as long as possible so that all possible squares are counted. For example, diagrams such as this are not considered valid.

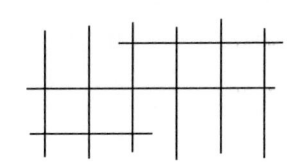

◊ With 9 lines 0, 6, 10 and 12 squares are possible:

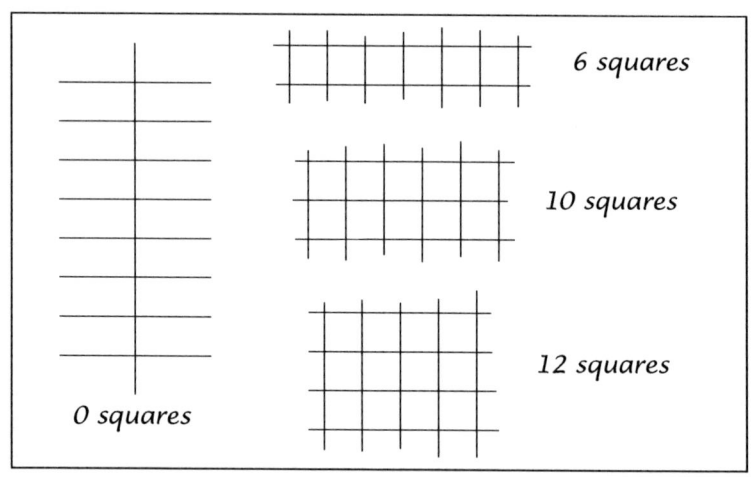

0 squares

6 squares

10 squares

12 squares

9 parallel lines will also produce 0 squares.

◊ All possible numbers of squares for 4 to 12 lines are

Number of lines	Numbers of squares
4	0 1
5	0 2
6	0 3 4
7	0 4 6
8	0 5 8 9
9	0 6 10 12
10	0 7 12 15 16
11	0 8 14 18 20
12	0 9 16 21 24 25

Pupils who produce a full set of results as in the table may make various observations such as:

- The minimum is always 0 and this can be achieved by a set of parallel lines.
- The next possible number of squares is $n - 3$ for n lines. These numbers go up by 1 each time.

◊ Encourage pupils to follow their own lines of investigation. For example, they could consider the maximum number of squares possible each time.

Encourage pupils to describe how the lines should be arranged to give the maximum number of squares by asking questions such as 'How would you arrange 100 lines to achieve the maximum number of squares?' and 'What about 99 lines?'

Formulas are

> even numbers of lines: $s = (\frac{n}{2} - 1)^2$
>
> odd numbers of lines: $s = (\frac{n}{2} - \frac{1}{2})(\frac{n}{2} - \frac{3}{2})$

Again, very few pupils are likely to express their conclusions algebraically at this stage.

B3 Swapover (p 305)

| Counters, tiles or pieces of paper |

◊ It is useful for pupils to draw large grids so the counters have plenty of room to move, especially diagonally.

◊ You may need to emphasise the starting position with the empty space on the right. The following starting position is not valid:

'Some found it frustrating at first. It helped if pupils worked in small groups.'

◊ Once pupils think they have found a solution to the puzzle on the page and are confident they can repeat it, they could think about ways of recording their results. Some will want to draw a diagram for each move. Encourage more able pupils to find efficient ways of presenting their results.

◊ Most pupils will tend to move the counters horizontally and vertically to start with. You may need to remind them they can move diagonally.

◊ 7 is the least number of moves for the grid shown.
The moves are

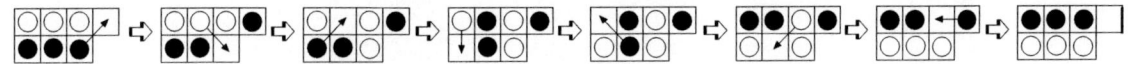

The method for the least number of moves can be described as always to move counters diagonally to the opposite side except where impossible. It can be easier to see this on larger grids.

◊ The minimum number of moves for various sizes of grid are

Counters in one line	Number of moves
1	3
2	5
3	7
4	9
5	11
6	13
7	15
8	17

Many pupils will notice that the number of moves goes up in 2s.
Some may be able to find that

number of moves = number of counters in one line × 2 + 1 or
number of moves = total number of counters + 1 or
number of moves = total number of squares

Ask pupils if they can use algebraic shorthand to write down their rules, stating clearly what their letters stand for.

Some may be able to show that their rule is correct for any sized grid. One year 8 boy put it nicely: 'The number of moves is the number of counters plus 1 because one of counters wastes a move by going sideways – all the others go up or down.'

◊ One possible extension is to only allow one diagonal move, the rest being horizontal or vertical. The method for the least number of moves in this case is harder to find and to describe. The rule is $n = 4c - 1$ where n is the number of moves and c is the number of counters in one line.
The initial moves for 4 counters in one line are

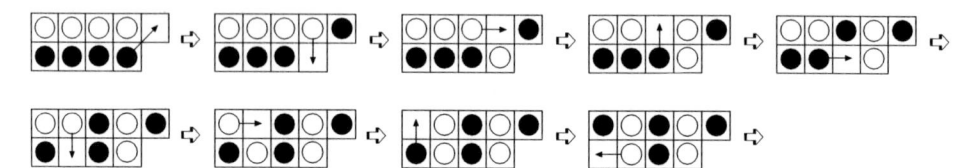

B4 Cutting a cake (p 305)

◊ Emphasise that each cut must go from one side of the cake to another.
For example, these cuts are not valid.

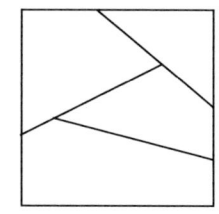

◊ Pupils could investigate the minimum and maximum number of pieces. The results are

Number of cuts	Minimum number of pieces	Maximum number of pieces
0	1	1
1	2	2
2	3	4
3	4	7
4	5	11
5	6	16
6	7	22
7	8	29
8	9	37

◊ Encourage less confident pupils to consider the minimum number of pieces first.

◊ Pupils may comment that:
- The minimum number of pieces goes up by 1 each time.
- The minimum number of pieces is always 1 more than the number of cuts (or, with n cuts, the minimum number of pieces is $n + 1$).
- The minimum number of pieces can be achieved by making a set of parallel cuts.
- The maximum numbers of pieces goes up by 1, then 2, then 3 and so on.

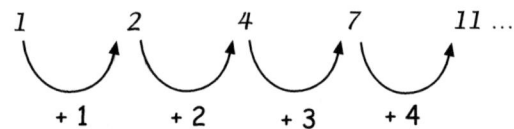

  1 → 2 (+1) → 4 (+2) → 7 (+3) → 11 (+4) ...

- To achieve the maximum number of pieces, take the previous diagram and make a cut that crosses each of the previous cuts.
- The sequence of numbers for the maximum number of pieces for one or more cuts (2, 4, 7, 11, 16, 22, ...) can be found by adding 1 to each of the numbers in the sequence for the maximum number of crossing points in *Crossing points* (1, 3, 6, 10, 15, 21, ...).

◊ Pupils may correctly predict the maximum number of pieces for various numbers of cuts but find it difficult to produce the corresponding diagrams. Using larger squares may help.

◊ For any number of cuts all numbers of pieces between the minimum and maximum can be achieved. Each diagram can be found by altering the previous one. For example, with four cuts:

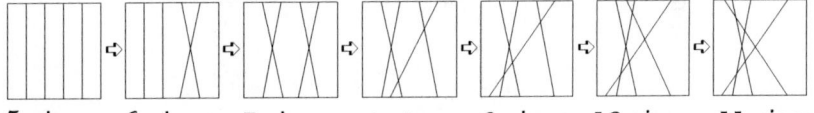

5 pieces 6 pieces 7 pieces 8 pieces 9 pieces 10 pieces 11 pieces

B5 Spots in a square (p 306)

◊ Emphasise that, if lines cross over, this has to be treated as creating more spots, for example

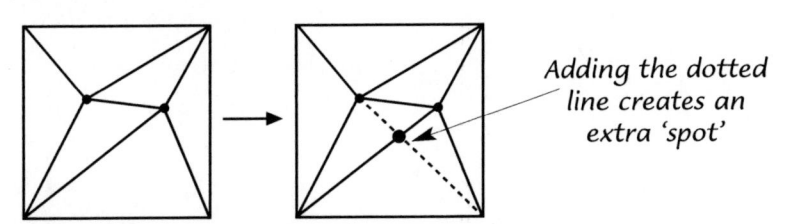

Adding the dotted line creates an extra 'spot'

42 Investigations • 231

◊ The results up to 6 spots (including 0 spots) are

Number of spots	Number of triangles
0	2
1	4
2	6
3	8
4	10
5	12
6	14

Many pupils will notice that the number of triangles goes up in 2s. Some may see that

number of triangles = number of spots × 2 + 2 or

number of triangles = (number of spots + 1) × 2

◊ Some pupils may realise that, given any diagram, you can add a spot in the middle of one of the triangles and join it to the three vertices of the triangles. This creates 3 smaller triangles inside the larger one, hence increasing the overall number of triangles by 2.

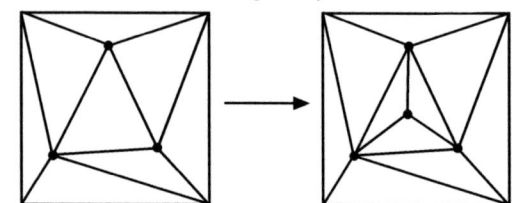

For example,

If a spot is added to an existing line, the two new lines will each add one to the number of the triangles.

B6 Turn, turn, turn (p 306)

This is a very rich activity with many possibilities for extension. Encourage pupils to follow their own lines of enquiry but some may need help in formulating questions.

Optional: Square dotty paper

◊ Many have found it beneficial for pupils to walk through the instructions to draw a turning track, emphasising the 90° turn each time.

◊ Encourage pupils to investigate questions such as:
- Do you always get back to your starting point?
  If you do, how many times do you repeat the instructions?
- What difference does it make if you turn left instead of right each time?
- What shape are the tracks?
  Do you get different types of tracks with different sets of numbers?

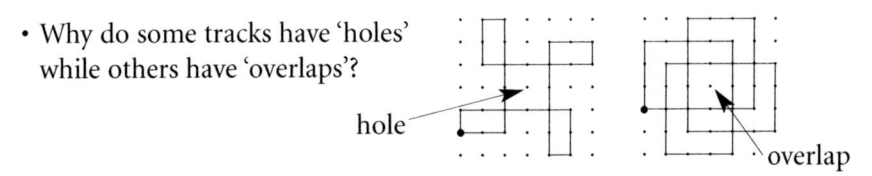

- Why do some tracks have 'holes' while others have 'overlaps'?

hole

overlap

Can you predict whether or not a track will have a hole or an overlap from the set of three numbers? Can you predict the size of the hole?

- What will happen with sets of consecutive numbers?
- What if you look at sets of numbers where the first two are always the same?
- What happens if you change the order of the numbers? Will the track for 1, 2, 4 look like the track for 4, 1, 2 for example?
- What if two or more of the numbers are the same?

◊ Pupils may notice these facts.

- It doesn't matter what order the numbers are in. You always produce the same track although it may be rotated or reflected.
- Turning left produces a reflection of the turning right track.
- If the numbers are all the same, you get a square track.
- If two of the numbers are the same, you get a cross shape.
- All tracks made with three numbers have rotation symmetry.
- If the two smallest numbers add up to the largest, you get a track with no hole or overlap.

  For example,

  1, 3, 2 (right turn)

  1, 3, 2 (left turn)

  7, 3, 4 (right turn)

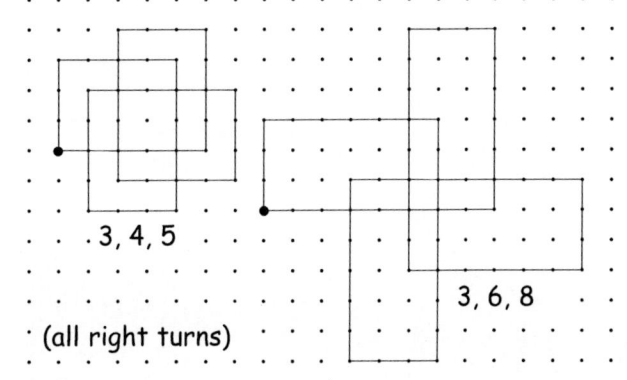

- If the sum of the two smallest numbers is greater than the largest number, you get a 'windmill' shape with an overlap.

  For example,

  3, 4, 5 (all right turns)

  3, 6, 8

42 Investigations

- If the sum of the two smallest numbers is smaller than the largest number, you get a windmill shape with a hole.
  For example,

  ·1, 5, 2
  (all right turns)

  · 3, 2, 6

  The size of the hole is the largest number minus the sum of the two smallest numbers.

◊ Investigating longer sets of numbers, pupils will find that 4 numbers produce infinite 'spiral' tracks and 5 numbers produce 'closed' tracks with rotation symmetry.
  For example,

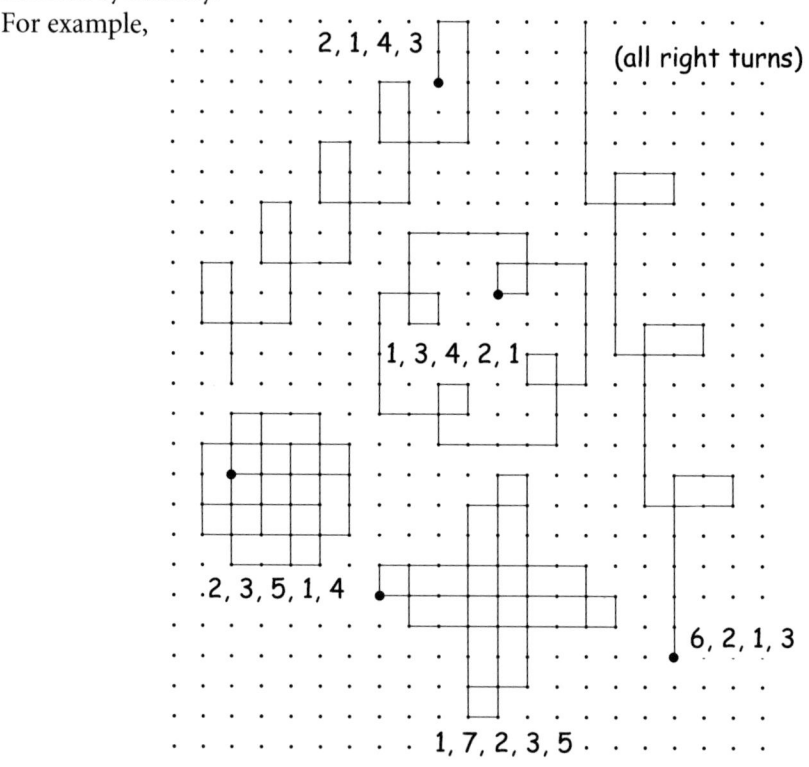

  2, 1, 4, 3   (all right turns)

  1, 3, 4, 2, 1

  ·2, 3, 5, 1, 4

  6, 2, 1, 3

  1, 7, 2, 3, 5

◊ Pupils can investigate what happens if you change the turning angle. For example, you could use a 60° right turn and investigate on triangular spotty paper.

◊ Pupils can use LOGO to draw their turning tracks.

43 Decimals with a calculator

Apart from any estimation, a calculator is to be used throughout.
Amounts of money are intended to be given correct to the nearest
penny throughout the unit.

p 307 **A** Mixed problems 1

p 308 **B** Multiplying by a decimal

p 311 **C** Dividing by a decimal

p 313 **D** Mixed problems 2

Practice booklet pages 95 to 98

A Mixed problems 1 (p 307)

B Multiplying by a decimal (p 308)

◊ You could start with some problems to be done mentally such as 'What is the cost of 2 metres of pure wool?'

Move on to problems such as 'What is the cost of 1.7 m of cotton?' where pupils should see that
- they need to calculate £9.90 × 1.7 to find the total cost
- they can estimate the cost by rounding the cost per metre and the length to the nearest whole number to give £10 × 2 = £20
- the cost lies between the cost of 1 m and the cost of 2 m

Include problems such as 'What is the cost of 2.45 m of pure wool?' that involve rounding to the nearest penny.

You could include discussion of the cost of, say, 0.8 m. (Multiplication by a number between 0 and 1 reduces.) This is covered in questions B11 onwards.

43 Decimals with a calculator • 235

C Dividing by a decimal (p 311)

◊ Encourage pupils who are not sure whether to multiply or divide, or which way round to divide, to try the problem with 'easier' numbers. For example, if 2 kg of ham cost £10, how could you work out how much 1 kg of ham would cost?

You should encourage pupils to check whether their answers seem sensible (though estimating divisions is not so straightforward as estimating multiplications).

D Mixed problems 2 (p 313)

A Mixed problems 1 (p 307)

A1 (a) 26.07 s (b) 2.57 s

A2 4.65 m

A3 £2.07

A4 (a) 1 kg, 3 kg, 1 kg, 1 kg, 2 kg
(b) 8 kg
(c) 7.81 kg

A5 Yes, the total length of the pieces of cable is 15.35 m which is less than 15.8 m.

A6 (a) £2 × 8 = £16 (b) £15.84

A7 £0.27 or 27p

A8 11.36 kg

B Multiplying by a decimal (p 308)

B1 (a) B (3 × 6) (b) £18.72

B2 (a) 6 × £13 = £78 (b) £80.52

B3 (a) 2 m of Irish linen will cost 2 × £12.50 = £25.
2.3 m is more than 2 m so will cost more than £25.
(b) £28.75

B4 (a) Estimate 4 × £8 = £32; £33.95
(b) Estimate 1 × £7 = £7; £8.70
(c) Estimate 4 × £8 = £32; £29.72
(d) Estimate 6 × £8 = £48; £52.16
(e) Estimate 2 × £13 = £26 or 2 × £12.50 = £25; £23.13
(f) Estimate 8 × £8 = £64; £65.30

B5 (a) 14 × 8 = 112 m^2
(b) 111.54 m^2

B6 £2.29

B7 (a) Less (b) £4.55

B8 (a) Blue costs £27.90.
White costs £25.58.
Red costs £22.79.
(b) £5.11

B9 £34.34; £32.14; £35.16

B10 (a)
1-litre	2.5-litre	5-litre	cost
7	1		£32.28
2	1	1	£28.08
2	3		£30.53

(b) The most sensible way is to buy two 5-litre cans and have some left over. This costs £26.50.

236 • 43 Decimals with a calculator

B11 (a) £14.89 (b) £48.75 (c) £3.25
(d) £3.40 (e) £1.02 (f) £2.07

B12 (a) 0.8 kg is less than 1 kg so the cost of 0.8 kg will be less than the cost of 1 kg which is £14.05.
(b) £11.24

B13 (a) 0.6 kg is between 0.5 kg and 1 kg so the cost of 0.6 kg will be between the cost of 0.5 kg and 1 kg, which is between £2.04 and £4.08.
(b) £2.45

B14 (a) 0.3 kg is less than 0.5 kg so the cost of 0.3 kg will be less than the cost of 0.5 kg which is £3.25.
(b) £1.95

B15 (a) £2.04 (b) £3.31 (c) £3.26
(d) £11.30 (e) £0.85 (f) £6.88

B16 £4.07 + £4.74 = £8.81

B17 £2.22 + £1.20 = £3.42

*B18 (a) £1.36 (b) £5.66 (c) £9.10

C Dividing by a decimal (p 311)

C1 (a) £0.57 (b) £0.60 (c) £0.65

C2 (a) 12 ÷ 1.5 (b) 8

C3 11

C4 15

C5 £2.30

C6 £0.48

C7 (a) £0.73 per litre
(b) £2.51 per kg
(c) £1.52 per metre
(d) £0.87 per kg

C8 (a) A: £2.84 per kg
B: £2.75 per kg
(b) Bag B gives more for your money as the cost per kg is less.

C9 (a) A: £1.67 per kg B: £1.33 per kg
Box B gives more for your money as the cost per kg is less.
(b) A: £1.30 per kg B: £1.23 per kg
Bag B gives more for your money as the cost per kg is less.
(c) A: £0.66 per kg B: £0.59 per kg
Bag B gives more for your money as the cost per kg is less.
(d) A: £0.94 per kg B: £1.15 per kg
Bag A gives more for your money as the cost per kg is less.

C10 £1.10

C11 (a) 1 kg is more than 0.3 kg so the cost of 1 kg will be more than the cost of 0.3 kg which is £0.60.
(b) £2.00

C12 (a) A: £11.96 per kg
B: £9.98 per kg
C: £8.99 per kg
D: £9.50 per kg
(b) Pack C (1 kg) is the best buy.
(c) Pupil's comments such as 'You might need less seed than this.'

D Mixed problems 2 (p 313)

D1 £0.72

D2 £8.25

D3 £5.52

D4 £1.05

D5 (a) A: £2.99 per kg B: £3.00 per kg
(b) Bag A is better value as the cost per kg is less.

D6 10.75 m

D7 No it is not enough. She needs £0.71 more.

D8 £1.02

D9 £3.87

D10 (a) The 15 kg bag (b) £12.70

D11 6.875 kg

D12 (a) 5   (b) £0.55

What progress have you made? (p 315)

1 (a) 3 × £8 = £24 (b) £24.77

2 (a) 2 kg of mushrooms will cost
2 × £2.50 = £5.
2.4 kg is more than 2 kg so will cost more than £5.
(b) £6.00

3 £2.15

4 £6.24

5 (a) £6.72   (b) £0.75
(c) £2.48   (d) £1.46

6 30

7 2.5-litre bottle: £0.62 per litre
1.5-litre bottle: £0.77 per litre
The 2.5-litre bottle gives you more for your money as it costs less per litre.

Practice booklet

Section A (p 95)

1 (a) 37.5 + 40.3 + 39.9 = 117.7 kg
(b) 40.3 − 37.5 = 2.8 kg

2 100 − 37.4 = 62.6 metres

3 200 × 9.54 = 1908 rand

4 £2.60 ÷ 3 = £0.87 (to the nearest penny)

5 Nikki 650 × 0.86 = 559 m
Pam 600 × 0.93 = 558 m
So Nikki is ahead by 1 m.

Section B (p 95)

1 (a) D (4 × 8) (b) £31.16

2 (a) The pupil's explanation, such as 'The cost of 3 kg of onions is over £1.50 so the greater weight of 3.2 kg will cost more too.'
(b) £1.66

3 (a) £1.96   (b) £1.30
(c) £0.63   (d) £0.97

4 (a) The pupil's explanation, such as 'The cost of 1 kg of tomatoes is £1.40 so the lesser weight of 0.8 kg will cost less than £1.40.'
(b) £1.12

5 (a) £0.68   (b) £0.24

Section C (p 96)

1 The larger piece costs £2.59 for 1 kg.
The smaller piece costs £2.37 for 1 kg.

2 25

3 £2.39

4 (a) £2.56 per kg (b) £4.60 per kg
(c) £0.98 per litre

5 (a) The smaller bag gives more for your money. The sugar in the smaller bag costs £1.08 per kg whereas the sugar in the larger bag costs £1.10 per kg.
(b) The larger container gives more for your money. The oil in the larger can costs £1.14 per litre whereas the oil in the smaller bottle costs £1.16 per litre.

Section D (p 97)

1 £0.43

2 (a) £4.54   (b) £1.34   (c) £16.55

3 £1.43

4 0.65 m

5 (a) £1.95
(b) Mrs Ling spends 5p more.

6 0.9575 kg

7 £14.20

44 Three dimensions

7C/37

Essential
Multilink cubes, triangular dotty paper

Optional
Sheet 170

Practice booklet pages 98 to 100

A Describing three-dimensional objects (p 316)

Multilink cubes (about 10 for each pair of pupils)

◊ The aim is to promote visualisation of three-dimensional shapes and the language needed to describe them.

The partner making the model from the description will almost certainly want to seek clarification and ask further questions. This can help pupils describe the shapes more precisely.

B Drawing three-dimensional objects (p 316)

Multilink cubes (about 6 per pupil), triangular dotty paper

C Views (p 317)

The choice of 'front' and 'side' view is arbitrary.

D Nets (p 319)

Optional: Sheet 170

◊ Pupils should be encouraged to do as much as they can by visualising the three-dimensional shapes: they should cut out and fold nets only when they cannot visualise the shapes, or to check their thinking.

◊ Nets P, Q and R are on the optional resource sheet 170. Having discussed their sketches, perhaps first in small groups and then as a whole class, pupils could make the solids.

◊ P square based pyramid   Q cube with corner sliced off   R 'house' or cube with triangular prism

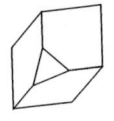

◊ Useful discussion questions are 'What is the same about the solid? What is different?' It is hoped that pupils will consider and discuss faces, vertices and edges, but do not force this.

E Surface area (p 321)

44 Three dimensions • 239

B Drawing three-dimensional objects (p 316)

B1 The pupil's drawing

B2 (a), (b), (c) The pupil's drawings
(d) T: 7   H: 11   E: 8

B3 If shapes that are 'mirror images' of each other are counted as different, 8 shapes can be made with four cubes.

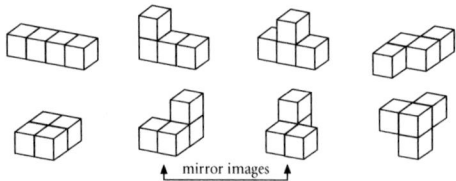

mirror images

B4 The pupil's drawings

C Views (p 317)

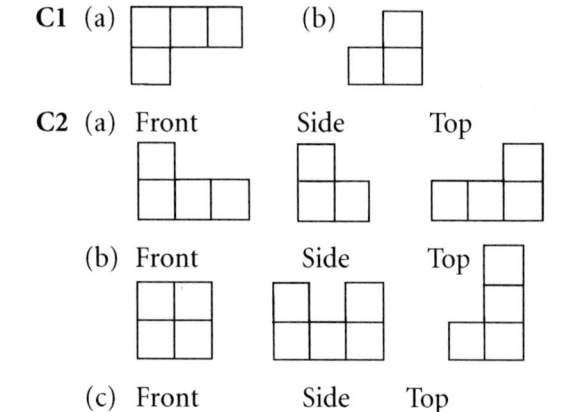

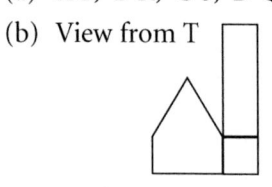

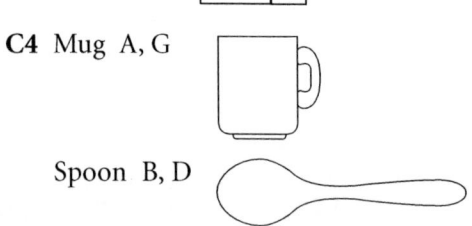

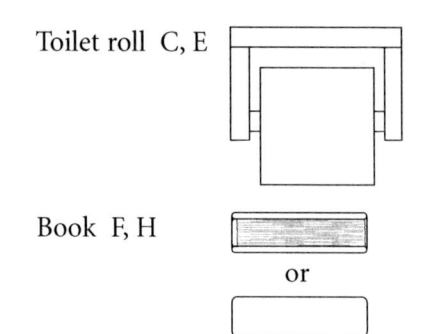

C1 (a)   (b)

C2 (a) Front   Side   Top
(b) Front   Side   Top
(c) Front   Side   Top

C3 (a) A P, B R, C S, D Q
(b) View from T

C4 Mug A, G

Spoon B, D

Toilet roll C, E

Book F, H

or

D Nets (p 319)

D1 (a) Could   (b) Could not
(c) Could not   (d) Could
(e) Could

D2 (a) The missing face could go on the net in any one of four positions.
(b)

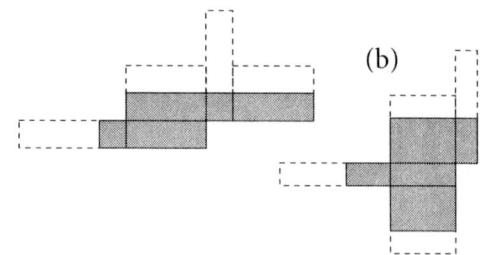

D3 (a) P and Q are nets of a cube, R is not.
(b) The arrangements which make a cube are

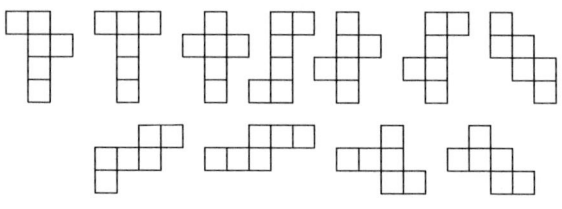

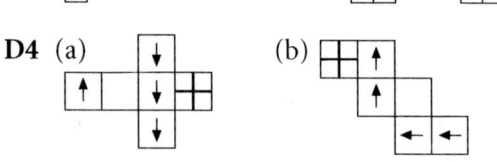

D4 (a)   (b)

D5 (a) Square based pyramid
(b) Triangular prism
(c) Tetrahedron

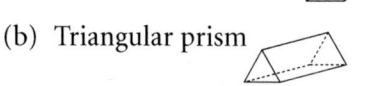

(d) Triangular prism ('wedge')

(e) Hexagonal prism

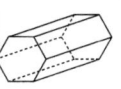

D6 (a) There are several possibilities. Here are two.

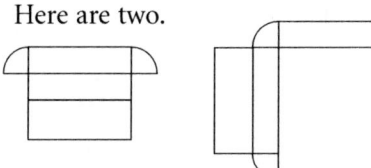

(b) There are several possibilities. Here is one.

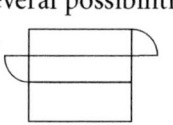

D7 This is one of several possibilities.

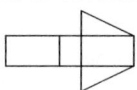

E Surface area (p 321)

E1 126 cm^2

E2 100 cm^2

E3 (a) 52 cm^2 (b) 122 cm^2
(c) 202 cm^2 (d) 158 cm^2

*****E4** 9 cm

What progress have you made? (p 321)

1 The pupil's model and drawings

2 This is one of several possibilities.

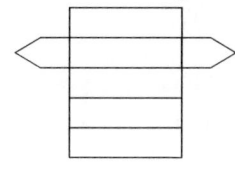

3 76 cm^2

Practice booklet

Sections B and C (p 98)

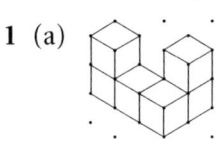

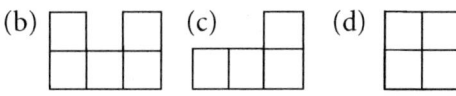

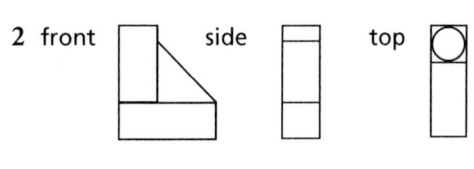

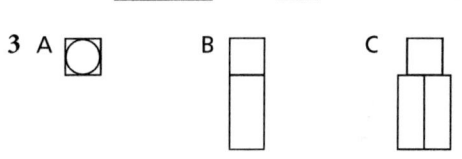

1 (a)

(b) (c) (d)

2 front   side   top

3 A   B   C

Section D (p 99)

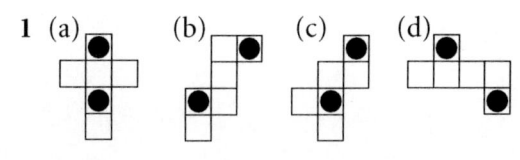

1 (a) (b) (c) (d)

2 B, E, F

Section E (p 100)

1 (a) 14 cm (b) 84 cm^2 (c) 104 cm^2

2 (a) 188 cm^2 (b) 136 cm^2 (c) 102 cm^2

3 7 cm

4 72 cm^2 (there are 18 squares)

㊺ Percentage

7C/31

This unit covers changing a percentage to a decimal, calculating a percentage of a quantity and expressing one quantity as a percentage of another. The latter leads on to drawing pie charts using a circular percentage scale.

> **Essential**
> Pie chart scale
> **Practice booklet** pages 101 to 105

Ⓐ Understanding percentages (p 322)

◊ Proportions are given in a variety of ways. From class or group discussion should emerge the need for a 'common currency' for expressing and comparing proportions.

◊ French cheeses may be labelled '40% matière grasse', for example; this means 40% of the dry matter is fat (i.e. the water content of the cheese is discounted).

◊ The order is:
Mascarpone 45%, Blue Stilton 36%, Red Leicester $33\frac{1}{3}$%, Danish Blue 28%, Edam 25%, Camembert 21%, Cottage Cheese 2%.

Ⓑ Percentages in your head (p 324)

Mental work with percentages can be returned to frequently in oral sessions.

Ⓒ Percentages and decimals (p 325)

Ⓓ Calculating a percentage of a quantity (p 327)

It is important that pupils should feel confident about their method of working out a percentage of a quantity. The 'one-step' method, treating the percentage as a decimal, is more sophisticated but ultimately better because it easily extends to a succession of percentage changes. It is used in section F.

Ⓔ Changing fractions to decimals (p 328)

You could point out that the division symbol ÷ is itself a fraction line with blanks above and below for numbers.

F One number as a percentage of another (p 329)

The approach used here depends on conversion from decimal to percentage.

G Drawing pie charts (p 331)

> Pie chart scale

◊ The pie charts shown are not labelled with the percentages, in order to give practice in measuring. However, it is a good practice to show the percentages.

◊ Rounding often leads to percentages which add up to slightly more or less than 100%. For this reason, when drawing pie charts it is often better to work to the nearest 0.1% as the small excess or deficit can be safely ignored.

A Understanding percentages (p 322)

A1 (a) About 27% (b) Water
(c) About 21%

A2 (a) C, D, F (b) G, H
(c) A, E (d) B, I

A3 (a) 20%–30% (b) 85%–95%
(c) 45%–55% (d) 55%–65%
(e) 3%–8% (f) 65%–75%

B Percentages in your head (p 324)

B1 (a) $\frac{1}{4}$ (b) $\frac{3}{4}$ (c) $\frac{1}{10}$
(d) $\frac{9}{10}$ (e) $\frac{1}{5}$

B2 (a) 50% (b) 10% (c) 25%
(d) 75% (e) $33\frac{1}{3}$%, 25%

B3 37%

B4 72%

B5 (a) £15 (b) £42 (c) £17.50

B6 (a) £10 (b) £21 (c) £17.50

B7 (a) To find 10%, you can divide by 10, giving 6 and 6.5.
(b) To find 5%, you can divide by 10, then halve.

B8 (a) 1p (b) 3p (c) 37p

B9 (a) 7p (b) 21p

C Percentages and decimals (p 325)

C1 (a) 0.5 (b) 0.25 (c) 0.65
(d) 0.78 (e) 0.1 (f) 0.01
(g) 0.04 (h) 0.4

C2 (a) 30% (b) 80% (c) 83%
(d) 3%

C3

Fraction	Decimal	Percentage
$\frac{70}{100}$ =	0.7 =	70%
$\frac{45}{100}$ =	0.45 =	45%
$\frac{57}{100}$ =	0.57 =	57%
$\frac{85}{100}$ =	0.85 =	85%
$\frac{5}{100}$ =	0.05 =	5%
$\frac{63}{100}$ =	0.63 =	63%
$\frac{7}{100}$ =	0.07 =	7%

45 Percentage • 243

C4 (a) 0.44 (b) 0.26 (c) 0.9
(d) 0.84 (e) 0.55 (f) 0.05
(g) 0.11 (h) 0.01 (i) 0.73
(j) 0.2 (k) 0.06 (l) 0.19

C5 (a) 72% (b) 31% (c) 50%
(d) 1% (e) 13% (f) 40%
(g) 2% (h) 92% (i) 4%
(j) 85% (k) 14% (l) 56%

C6 1%, 0.1, $\frac{12}{100}$, 15%, 0.25, 0.3, $\frac{45}{100}$

C7 $\frac{78}{100}$, 0.75, 65%, 0.51, 0.4, 0.08, 5%

C8 (a) (i) $\frac{30}{100}$ (ii) 30%
(b) (i) $\frac{40}{100}$ (ii) 40%
(c) (i) $\frac{70}{100}$ (ii) 70%
(d) (i) $\frac{20}{100}$ (ii) 20%
(e) (i) $\frac{60}{100}$ (ii) 60%
(f) (i) $\frac{15}{100}$ (ii) 15%
(g) (i) $\frac{65}{100}$ (ii) 65%
(h) (i) $\frac{35}{100}$ (ii) 35%

C9 0.08, $\frac{1}{10}$, 15%, 0.2, 0.56, 74%, $\frac{3}{4}$

D Calculating a percentage of a quantity (p 327)

D1 (a) 162 g (b) 256.2 g (c) 66.7 g
(d) 93.8 g (e) 26.6 g (f) 35.2 g
(g) 201.6 g (h) 52.8 g

D2 (a) 2.25 g (b) 6.72 g
(c) 3.6 g (d) 1.68 g

D3 410 g

D4 0.6 g

D5 He is right for 10%, but dividing by 5 gives 20%, not 5%.

D6 (a) Sugar 19.95 g, fat 10.5 g, protein 2.8 g
(b) Sugar 85.5 g, fat 45 g, protein 12 g
(c) Sugar 285 g, fat 150 g, protein 40 g

E Changing fractions to decimals (p 328)

E1 (a) 0.25 (b) 0.125 (c) 0.05
(d) 0.8 (e) 0.375 (f) 0.875
(g) 0.28 (h) 0.15 (i) 0.22
(j) 0.9375

E2 $\frac{29}{50}$ (0.58), $\frac{3}{5}$ (0.6), $\frac{5}{8}$ (0.625), $\frac{13}{20}$ (0.65)

E3 $\frac{17}{20}$ (0.85), $\frac{39}{50}$ (0.78), $\frac{19}{25}$ (0.76), $\frac{3}{4}$ (0.75)

E4 (a) 0.14 (b) 0.57 (c) 0.11
(d) 0.56 (e) 0.64 (f) 0.27
(g) 0.08 (h) 0.38 (i) 0.41
(j) 0.87

F One number as a percentage of another (p 329)

F1 (a) $\frac{5}{7}$ (b) 71% (to the nearest 1%)

F2 (a) 60% (b) 35% (c) 87.5%
(d) about 33% (e) about 67%

F3 (a) 29% (b) 78% (c) 23%
(d) 41% (e) 5%

F4 (a) $\frac{1}{4}$ (b) 25% (c) 19%

F5 (a) 30% (b) 70%

F6 (a) 86% (b) 14%

F7 30%

F8 Peter attended 83% of the practices. Carol attended 78% of the practices. So Peter has the better record.

F9 (a) Model A (3% faulty)
(b) Model D (14% faulty)

*F10 (a) 16%
(b) In a class of 30 it would be about 5 people.

G Drawing pie charts (p 331)

G1 These features (among others) may be noticed: Cheese spread has higher proportions of water and carbohydrate, but lower proportions of fat and protein.

G2 (a) 28% (b) 23% (c) 46% (d) 53%

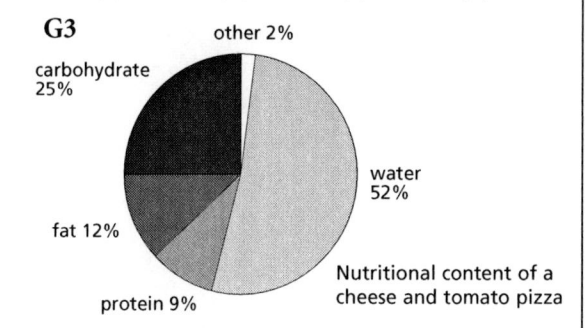

G3 Nutritional content of a cheese and tomato pizza:
carbohydrate 25%, other 2%, water 52%, fat 12%, protein 9%

G4 (a) Meat, fish and eggs (b) 9%
(c) Clio is wrong. The chart shows money spent, not the amounts eaten.

G5 sugar, sweets, jams, fats, oils 5%; milk, butter, dairy products 10%; bread, cakes, cereals 25%; meat, fish, eggs 35%; fruit 10%; vegetables 15%

G6 (a) 32 pages (b) 34%
(c) Foreign news 22%, sport 16%, entertainment 9%, finance 19%

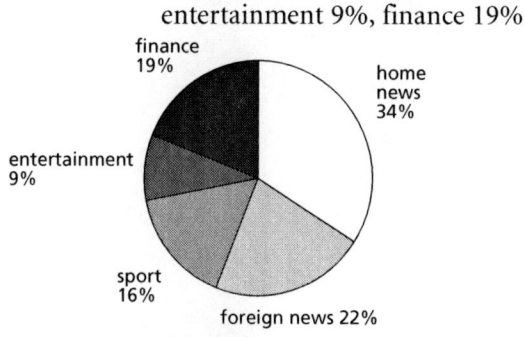

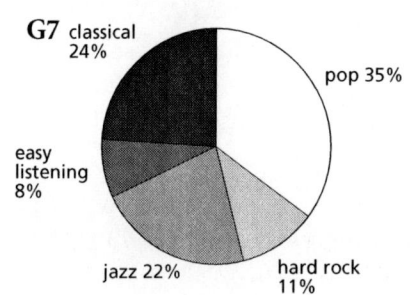

finance 19%, home news 34%, entertainment 9%, sport 16%, foreign news 22%

G7 classical 24%, pop 35%, easy listening 8%, jazz 22%, hard rock 11%

What progress have you made? (p 333)

1 (a) C (b) A (c) E
2 (a) 0.5 (b) 0.45 (c) 0.04 (d) 0.07
3 (a) £5 (b) 2 kg (c) 9 kg (d) £2
4 (a) 68.4 g (b) 13 g
5 73%
6 21 out of 25 (84%) is better than 30 out of 37 (81%).
7 draws 25%, wins 60%, losses 15%

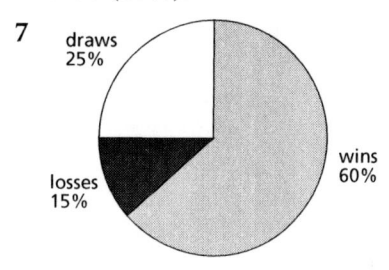

Practice booklet

Section A (p 101)

1 (a) 30% (b) 10% (c) 70% (d) 50%

Section B (p 101)

1 (a) 25% (b) 30% (c) 20%
2 55%
3 (a) £10 (b) £8.50 (c) £45
  (d) £15 (e) £13 (f) £12.50
  (g) 7 kg (h) £7.50 (i) £7.30
4 (a) 5p (b) 15p (c) 90p
  (d) £4.50
5 (a) 3% of £10 is £0.30. Others are £3.
  (b) 20% of £30 is £6. Others are £5.
  (c) 5% of £80 is £4. Others are £4.50.

Section C (p 102)

1 (a) 0.68 (b) 0.41 (c) 0.8
  (d) 0.08 (e) 0.09 (f) 0.9
  (g) 0.54 (h) 0.3

2 (a) 65% (b) 60% (c) 6%
  (d) 7%   (e) 2%   (f) 22%
  (g) 20% (h) 3%

3 (a) 21% (b) 40% (c) 45% (d) 80%

4 (a) 7% $\frac{8}{100}$ 0.09   0.1   0.8 $\frac{81}{100}$

  (b) 0.02   0.1   12%   20% $\frac{1}{2}$ 0.7

  (c) 0.07 $\frac{9}{100}$ 0.3 and $\frac{3}{10}$ 40%   75%

Section D (p 103)

1 (a) 3.2 kg   (b) 3.3 kg   (c) 3.3 kg
  (d) 1.12 kg (e) 103.4 kg (f) 21.32 kg

2 (a) £1530   (b) £6970

3   34 g

4 (a) £410   (b) £36 306   (c) £2856

Sections E and F (p 103)

1 (a) $\frac{6}{11} = 55\%$, $\frac{11}{19} = 58\%$, $\frac{13}{27} = 48\%$

  (b) $\frac{13}{27}$, $\frac{6}{11}$, $\frac{11}{19}$

2   76%

3   12%

4   31%

5 (a) Screen 3 (87%)
  (b) Screen 2 (37%)

Section G (p 104)

1 (a) The pupil's comments, such as 'The proportion working in agriculture is the same in both places.
A bigger proportion are working in service industries and in distribution and catering in Latin America.
A bigger proportion are working in manufacturing, construction and transport and communications in Eastern Europe.'

  (b) Eastern Europe   23%
      Latin America   33%

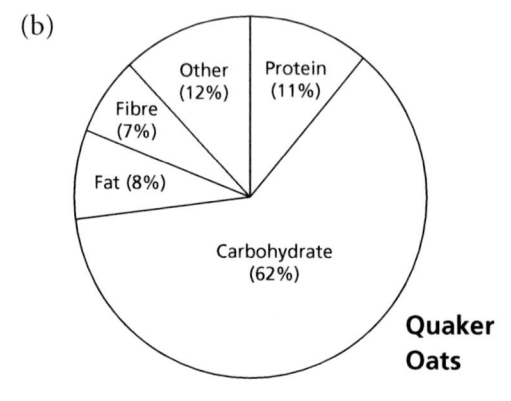

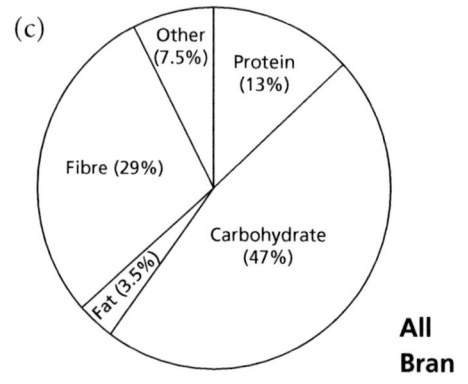

2 (a) 80 g
  (b) Quaker Oats
      Other (12%)
      Protein (11%)
      Fibre (7%)
      Fat (8%)
      Carbohydrate (62%)

  (c) All Bran
      Other (7.5%)
      Protein (13%)
      Fibre (29%)
      Fat (3.5%)
      Carbohydrate (47%)

  (d) The protein content of each is similar, but All Bran contains a lot more fibre and less carbohydrate and fat than Quaker Oats.

Review 5 (p 334)

Essential
Graph paper, triangular dotty paper, calculator

1. (a) $2n + 9 = 23$, $n = 7$
   (b) $4(n - 17) = 16$, $n = 21$
   (c) $6n - 15 = 0$, $n = 2.5$

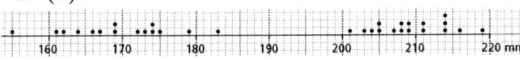

2. (a) [number line from 160 to 220 mm with data points]
   (b) 64 mm   (c) 202 mm
   (d) The handspans are in two clusters.

3. (a) D   (b) C   (c) A   (d) B

4. (a) £11.34   (b) £0.42 or 42p

5. Star

6. (a) The pupil's copy of the shape with the centre of rotation marked half-way between C and G
   (b) 2
   (c) No
   (d) FE
   (e) AH and DE
   (f) The pupil's two right angles marked on the shape ($\angle$BAH and $\angle$DEF are interior right angles, $\angle$BCD and $\angle$HGF are also right angles)

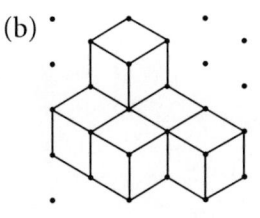

7. $\frac{1}{2}$ 50%
   $\frac{1}{4}$ 25%
   $\frac{1}{5}$ 20%
   $\frac{3}{5}$ 60%
   $\frac{3}{10}$ 30%
   $\frac{3}{4}$ 75%

8. (a) B   (b) [isometric drawing of cubes on triangular dotty paper]

9. (a) 0.34 kg   (b) 3.06 kg

10. 23 days

11. £24

12. 0.03, $\frac{1}{4}$, 30%, 0.42, $\frac{4}{5}$, 90%

13. (a) 5928   (b) 6000

14. (a) A: Line 2   B: Lines 1 and 3
    (b) A: Order 2   B: Order 3

15. (a) One correct net is [diagram of net with dimensions 5.3 cm, 4.1 cm, 12.8 cm]
    (b) 284.1 cm²

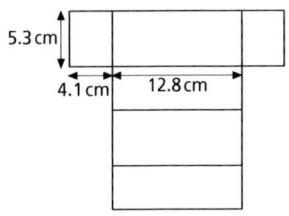

16. (a) £30   (b) £27.89

17. 14.7 kg

18. (a) $z = 12$ (b) $p = 1$
    (c) $d = 5.8$ (d) $x = 8.5$

19. (a) 13%   (b) 0.120 kg

20. £14.99

Mixed questions 5 (Practice booklet p 105)

1. (a) (i) 36   (ii) 37
   (b) (i) 50   (ii) 36
   (c) (i) T   (ii) F   (iii) F

2. (a) 75%   (b) 25%

3. (a) 2400   (b) 2400
   (c) 2000   (d) 14 000

4. $2n + 5 = 19$, $n = 7$

5. (a) 3   (b) 15   (c) 2   (d) 2

6. (a) 832   (b) 21

7

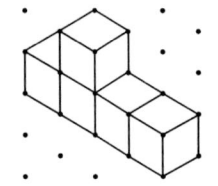

8 The 2.5 kg bag gives you more for your money: these apples cost 36p per kg but the apples in the smaller bag cost 37p per kg.

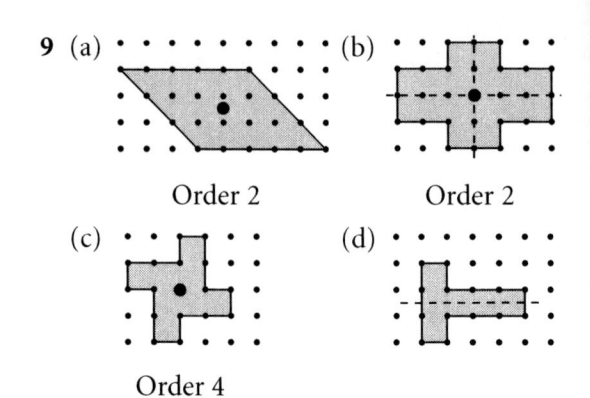

9 (a) Order 2   (b) Order 2

 (c) Order 4

 (d)

10 (a) $q = 7$ (b) $r = 8$ (c) $x = 24$

11 39.59 zloty

12 69%

46 Fair to all?   7C/35

p 337 **A** How to be fair	Finding the mean Comparing two sets of data using means
p 340 **B** Means from frequencies	Finding the mean from a frequency table
p 344 **C** Averages	Comparing using mean, median and mode

Essential	**Optional**
Packs of playing cards or number cards (for 'Mean tricks') Newspapers/foreign language texts (for the investigation on page 343)	Sheet 171

Practice booklet pages 107 to 110

A How to be fair (p 337)

Decimal means occur in this section but no rounding is necessary.

> Optional: Sheet 171

◊ You could start by asking the class which group they think did better.

Ann's group collected 20 kg of paper, Ben's group collected 18 kg. However, there are more pupils in Ann's group, and pupils may argue that Ben's group collected 6 kg per pupil while Ann's group only managed 5 kg per pupil. Introduce the mean weight of newspapers as the result of sharing the newspaper equally between members of the group.

Sheet 171 can be used to illustrate this further.

Pupils should be aware that the existence of a mean value does not imply that such a value must exist in practice. For example, the mean for Ann's group is 5 kg and no one collected that amount.

You may wish to extend your initial discussion to include decimal means. For example, 'What if Ann collected 5 kg instead of 3 kg?'

◊ Decimal means first occur in question A7. You may wish pupils to try A1 to A6 without a calculator to consolidate number skills. Use of a calculator is assumed in all subsequent work.

A7 This can generate discussion. Using means puts Wayne (8.5) at the top followed by Pat (8) and Jon (7.8). However, using medians puts Jon (8) at the top followed by Pat (7.5) and Wayne (6.5). Pupils may also bring in the idea of consistency.

46 Fair to all? • 249

Mean tricks (p 340)

'We played as a class divided into two halves – works well. Worth persevering – really shows if they understand the concept.'

Pupils can practise their mental skills in calculating the mean for small data sets and the game also helps them consolidate the idea of the mean.

> Packs of playing cards (picture cards removed). Alternatively, sets of 40 number cards (four 1s, four 2s, four 3s etc.) could be used instead.

◊ Schools have found that the benefits of this game improve with playing as pupils develop strategies to calculate the mean quickly.

Suppose the target mean is 6. Some pupils may use the strategy of looking for two cards with a total of 12, three cards with a total of 18, four cards with a total of 24 and so on. Others may use a method where they try to make the total deviation from the mean zero. For example the total deviation of 5, 4, 6 and 9 from 6 is $^-1 + {}^-2 + 0 + 3 = 0$ so the mean of this set must be 6. Pupils could compare their methods.

B **Means from frequencies** (p 340)

Pupils need to be able to round to one or two decimal places.

◊ Show how the frequency table relates to the weights of the players in the football team. Pupils should be clear that, say, the first line of the table shows that there are 3 people that weigh 70 kg in the team.

Emphasise that pupils are to try to answer the questions just by looking at the numbers in the table. However, looking at the illustration above the table should help clarify things for pupils who find this difficult.

◊ It is worth spending time on the meaning of the word 'frequency' with respect to each question. It is a very common misconception to add up the frequencies and divide that by the number of frequencies that occur.

◊ Encourage pupils to lay out their work carefully. It may help them to add a 'total' column to each table and fill this in. The table on page 340 could be extended like this

Weight in kg	Frequency	Total weight
70	3	210
71	2	142
72	4	288
73	2	146

When working with frequency tables it can help trace a mistake if all steps are shown.

B12 In part (c), watch out for pupils who try to find the mean age of the choir by finding the mean of the means found in parts (a) and (b), obtaining 13.625 years.

◊ An additional activity is to choose a book and try to estimate the number of words in the book. This is likely to involve estimating the mean number of words on a line and working from there. Pupils could also consider questions such as 'How many pages in the book would come to 5000 words?'

Investigation (p 343)

> Newspapers and foreign language texts

◊ Pupils can of course compare any two papers; collated results make an interesting wall display. Samples should be of a reasonable size – counting the number of words in, say, 40 sentences and finding the mean using a frequency table would be fine.

This has been found to be a very useful homework task.

C **Averages** (p 344)

This section brings together the three measures of average – mean, median and mode – and introduces the idea of the most representative average.

◊ Discussion of the three initial statements should bring out that the word 'average' is used loosely to mean 'about the middle'.

In the statement 'Mary is of average height.' Is 'average height' the average for the school year? the school? the population of the UK?

What average could it be?

What average do you think is appropriate? Why?

The other statements can be discussed similarly.

◊ Pupils should think about which of the pocket money averages is a good representative average for this data. They are likely to find this idea quite difficult. They may see that the mode is not very representative as over half the young people receive £4.50 or more. Choosing between the mean and median is more difficult. Ask pupils if the median would change if the five pupils that received £3.00 or less actually all received £1.00. This may help them see that the median is unchanged by the relative poverty of these five! In this case the mean can be regarded as the most representative as there are no very large or small values to distort it.

◊ In questions C2 to C4, ordering the data can be tedious. Encourage pupils to think about how to record the results to make finding the averages easier, for example a dot plot or tally chart/frequency table.

A How to be fair (p 337)

A1 (a) 6 kg (b) 7 kg (c) Jason's group

A2 27 ÷ 9 = 3

A3 120 ÷ 10 = 12 goldfish

A4 (a) 40 ÷ 5 = 8 pupils
(b) 36 ÷ 4 = 9 pupils

A5 35 ÷ 5 = 7 peppers per plant

A6 Ruth is right: the 24 peppers have to be shared between the four plants.

A7 Pat 32 ÷ 4 or 8 points per game
Jon 39 ÷ 5 or 7.8 points per game
Wayne 51 ÷ 6 or 8.5 points per game

You could argue that Wayne is the best points scorer as his mean number of points is the highest. However, it could also be argued that Pat's mean is only 0.5 less than Wayne's and she is more consistent that him.

A8 (a) Means: 8L 40 ÷ 20 = 2 cans
8N 55 ÷ 25 = 2.2 cans

8N has a higher mean and you could argue from this that it should get the prize.

(b) Means: 8L 50 ÷ 20 = 2.5 bottles
8N 60 ÷ 25 = 2.4 bottles

8L has a higher mean and you could argue from this that it should get the prize.

(c) The pupil's view and explanation
8L collected a mean of 90 ÷ 20 = 4.5 items per person and 8N collected a mean of 115 ÷ 25 = 4.6 items per person so you could argue that 8N did better overall.

A9 Answers can be found by calculating the mean load pulled by a member of the team (although this assumes a similar level of fitness for each team member).

The mean loads pulled per person are:
Team A: 2475 ÷ 9 = 275 kg
Team B: 1656 ÷ 6 = 276 kg
Team C: 1911 ÷ 7 = 273 kg
(a) Team B (b) Team C
The pupil's reasons

B Means from frequencies (p 340)

B1 (a) 15 players
(b) 5 players
(c) (75 × 1) + (76 × 4) + (77 × 3) + (78 × 5) + (79 × 2) = 1158 kg
(d) 1158 ÷ 15 = 77.2 kg

B2 (a) 35 nests
(b) (3 × 17) + (4 × 15) + (5 × 3) = 126 eggs
(c) 126 ÷ 35 = 3.6 eggs

B3 (a) 40 games
(b) 23 games
(c) (0 × 5) + (1 × 12) + (2 × 13) + (3 × 8) + (4 × 2) = 70 goals
(d) 70 ÷ 40 = 1.75 goals

B4 (a) 40 plants
(b) (4 × 7) + (5 × 9) + (6 × 11) + (7 × 10) + (8 × 3) = 233 tomatoes
(c) 233 ÷ 40 = 5.8 tomatoes (to 1 d.p.)

B5 (26 × 2) + (28 × 22) + (29 × 13) + (30 × 13) = 1435 Smarties
1435 ÷ 50 = 28.7 Smarties

B6 (a)

Price	Frequency
57p	3
58p	5
59p	5
60p	3
61p	4

(b) (57 × 3) + (58 × 5) + (59 × 5) + (60 × 3) + (61 × 4) = 1180p
1180 ÷ 20 = 59p

B7 5.97 (to 2 d.p.) with the pupil's description

B8 (8 × 2) + (7 × 3) + (6 × 9) + (5 × 1) = 96
96 ÷ 15 = 6.4 tomatoes

B9 (1 × 40) + (2 × 30) + (3 × 18) + (4 × 12) + (5 × 4) = 222 people
222 ÷ 104 = 2.1 people (to 1 d.p.)

B10 (a) (48 × 10) + (49 × 27) + (50 × 39) + (51 × 67) + (52 × 34)
= 8938 matches
8938 ÷ 177 = 50.5 matches (to 1 d.p.)

(b) The statement seems fair. Although you might only get 48 matches, the mean is more than the 50 stated.

B11 (a) The pupil's set of five numbers
(b) The mean increases by 2.
(c) The mean increases by the number you add to each of your set of numbers.
(d) 9760 + [(1 + 3 + 0 + 2 + 4) ÷ 5] = 9762

B12 (a) 265 ÷ 20 = 13.25 years
(b) 420 ÷ 30 = 14 years
(c) 685 ÷ 50 = 13.7 years

B13 (a) 11.75 years (b) 14 years

C **Averages** (p 344)

C1 (a) Median (b) Mode

C2 (a) £2.50 (b) £2.35
(c) £3.00 (d) Mode

C3 (a) (i) 23 years
(ii) 794 ÷ 30 = 26.5 years (to 1 d.p.)
(iii) 18 years

(b) It could change at a day's notice, depending on when members' birthdays are. The frequencies are also too low for the mode to be relevant.

(c)
Age	Frequency
10–19	11
20–29	9
30–39	6
40–49	3
50–59	1

The modal age group is 10–19.

C4 (a)
Hours	Number of days (Blackmouth)	Number of days (Bournepool)
1	2	3
2	1	1
3	3	2
4	5	4
5	5	3
6	2	2
7	1	7
8	3	6
9	4	3
10	3	0
11	2	0

Blackmouth
median: 5 hours
mean: 189 ÷ 31 = 6.1 hours (to 1 d.p.)
modes: 4 and 5 hours

Bournepool
median: 7 hours
mean: 178 ÷ 31 = 5.7 hours (to 1 d.p.)
mode: 7 hours

Although the mean number of hours of sunshine is less for Bournepool (5.7 compared with 6.1), the median is higher. This means that half the days at Bournepool have 7 or more hours of sunshine compared with 5 or more at Blackmouth. The mode of 7 hours at Bournepool is more than that of the two most frequent amounts of sunshine at Blackmouth (4 and 5 hours).

(b) The pupil's adverts

Challenge

It can take some time to list them all, so you could just ask for what pupils can find.

There are 21 sets, some of which are listed here:
1, 2, 3, 5, 6, 18, 21   1, 2, 3, 5, 11, 13, 21
1, 2, 4, 5, 6, 17, 21   1, 3, 4, 5, 8, 14, 21
2, 3, 4, 5, 9, 11, 22

What progress have you made? (p 346)

1. £37.24 ÷ 7 = £5.32

2. (a) Mean: 137 ÷ 30 = 4.6 hours (to 1 d.p.)
   Mode: 0 hours
   Median: 5 hours

   (b) Median or mean. The mode of 0 hours is not at all representative.

3. 1155 ÷ 50 = 23.1 calculators

Practice booklet

Section A (p 107)

1. 14 ÷ 7 = 2 hours

2. (a) The girls' mean pocket money is £40.00 ÷ 10 = £4.00
   (b) The boys' mean pocket money is £31.60 ÷ 8 = £3.95
   (c) The girls did better.

3. Kathy did better.
   Her mean score was 3.5, higher than Mark's mean score of 3.4.

4. (a) The pupil's answer, for example 5, 7, 9 or 3, 5, 13
   (b) The total of the three numbers must be 21 which is an odd number. However, when even numbers are added the answer is always even.

5. The pupil's answer (the total of the five numbers must be 47)

Section B (p 108)

1. (a) 100
   (b) (1 × 50) + (2 × 25) + (3 × 20) + (4 × 3) + (5 × 2) = 182
   (c) 182 ÷ 100 = 1.82 people

2. (4 × 5) + (5 × 1) + ... + (14 × 4) = 460
   460 ÷ 50 = 9.2 eggs

Section C (p 109)

1. (a) 5410 ÷ 25 = 216.4 grams
   (b), (e) and (g)

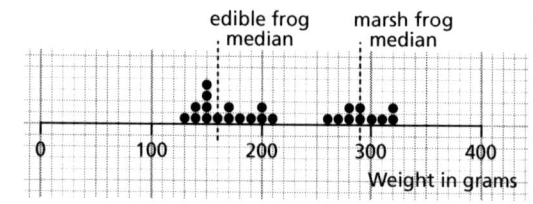

   (c) 10
   (d) 2920 ÷ 10 = 292 grams
   (e) Median of 290 g marked on dot plot
   (f) 2490 ÷ 15 = 166 grams
   (g) Median of 160 g marked on dot plot
   (h) The pupil's answer

2. (a) 169.5 cm
   (b) 5096 ÷ 30 = 169.9 cm (to 1 d.p.)
   (c) 173 cm
   (d) The pupil's explanation: for example, only $\frac{1}{10}$ of the pupils are above the modal height.

3. (a) 63 ÷ 49 = 1.29 bottles per home
   (b) 0   (c) 1
   (d) 63 ÷ 26 = 2.42 bottles

*4 The pupil's set of numbers, for example 0, 1, 9, 10, 10

*5 The pupil's set of numbers, for example 0, 1, 4, 10, 10

47 Transformations

Essential	Optional
Sheets 172 and 173	OHP
Tracing paper	Transparency of sheet 172
Squared paper	Small mirrors

Practice booklet pages 111 and 112

A Reflection (p 347)

> Sheet 172, tracing paper, squared paper
> Optional: Transparency of sheet 172, small mirrors

◊ Pupils can experiment with the picture of Dinah, its images and mirror lines on the top part of sheet 172, using tracing paper or mirrors to check the reflections where necessary. You can lead this work using a transparency of this sheet. If pupils use tracing paper, it is a good idea if they first put an arrow head on the mirror line they are using and trace it when they trace the shape: then the tracing paper can be put in the correct position when it is turned over. An aim of the work is for pupils gradually to reduce their dependence on tracing paper or a mirror, though of course these aids are always useful for checking.

Here and in the later sections, use – and encourage the use of – language like 'the original shape', 'the image after a reflection' and so on.

Negative coordinates are gradually introduced in this section and the next. You may need to do extra practice on plotting and identifying points with negative coordinates.

B Rotation (p 349)

> Sheets 172 and 173
> Tracing paper, squared paper

Pupils can work on the lower part of sheet 172, which reproduces the quadrilateral, and you can also use a transparency of this sheet to lead the work. Some of these questions may arise as you do so:

- How do I know when I've completed a half turn?
- How many degrees are there in a half turn?
- How many in a quarter turn?

The importance of giving a full description of a rotation (centre, angle, direction) should be stressed.

◊ The rotations in B2(c) (on sheet 173) are quite challenging, as the position of each centre of rotation is not obvious.

C Translation (p 352)

> Squared paper

◊ Karl's error in C2 shows the need to concentrate on what happens to a particular point on the starting shape (in this case the bottom right-hand corner of the triangle) when a translation occurs.

◊ Column vector notation is not used at this stage but you could introduce it now if you wish.

◊ In introducing the 'Patterns with transformations' activity to pupils, you may need to stress that they should describe what needs to be done to the *original design* to get each part of the pattern, not what needs to be done to the part of the pattern they have just dealt with.

A Reflection (p 347)

A1

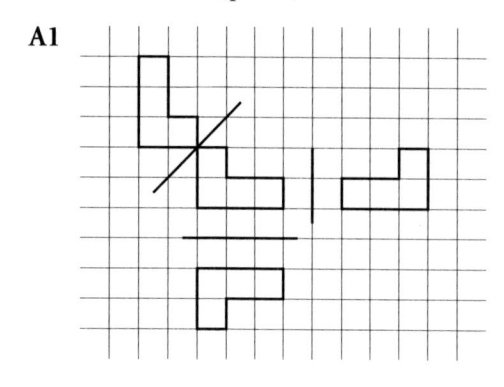

A2 (a) V (b) X (c) W (d) U

A3 (a) E (b) C (c) G (d) J

A4 (a) M4 (b) M5 (c) M1 (d) M1

A5

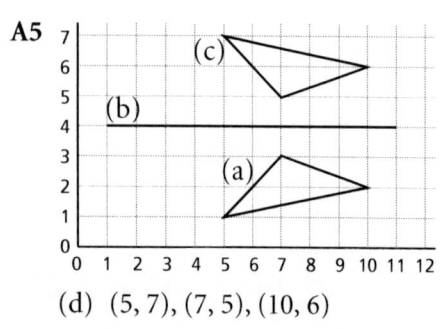

(d) (5, 7), (7, 5), (10, 6)

A6

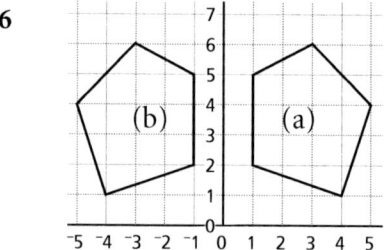

(c) (⁻1, 2), (⁻4, 1), (⁻5, 4), (⁻3, 6), (⁻1, 5)
In each pair of coordinates the first (across) number has become negative.

B Rotation (p 349)

B1
90° clockwise about C
90° anticlockwise about C
90° anticlockwise about E
180° about D

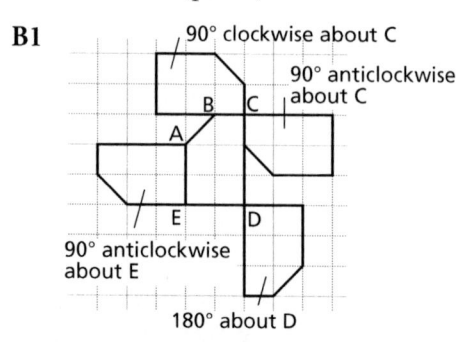

B2 The pupil's rotations on sheet 173

B3 (a) W (b) X (c) Z (d) Y (e) U

B4 (a) Centre A, 180°
(b) Centre C, 90° clockwise
(c) Centre B, 180°
(d) Centre B, 180°

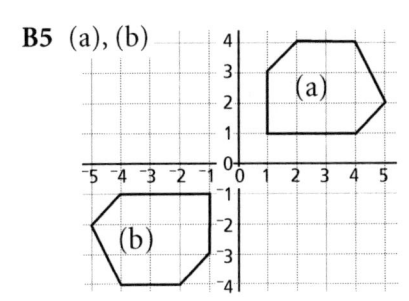

B5 (a), (b)

(c) (⁻1, ⁻1), (⁻1, ⁻3), (⁻2, ⁻4), (⁻4, ⁻4), (⁻5, ⁻2), (⁻4, ⁻1)
All the coordinates (up and across) have become negative.

Transformation code

The two words are STONE and NOSE.

For TASTE, starting at T:
Reflect in M4
Reflect in M2
Rotate a half turn, centre C
Reflect in M2

For ROTATE, starting at R:
Rotate a half turn, centre C
Rotate a quarter turn clockwise, centre C
Reflect in M4
Reflect in M4
Reflect in M2

C Translation (p 352)

C1 (a) E (b) B (c) C (d) G

C2 He is not correct; the translation is 4 units right, 1 unit up.

C3 (a) 2 units right, 1 unit up
(b) 2 units left, 1 unit up
(c) 4 units left, 2 units down
(d) 4 units right, 2 units down

C4 (a)

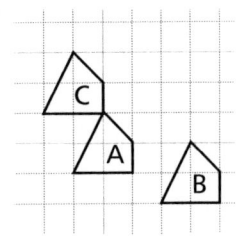

(b) 4 units left, 3 units up
(c) 4 units right, 3 units down

Patterns with transformations

1 (a) Reflect design in line M4
(b) Rotate design 180° about O
(c) Reflect design in M2

2 (a) Rotate design 90° clockwise about O
(b) Rotate design 180° about O
(c) Rotate design 90° anticlockwise about O

3 (a) Rotate design 90° clockwise about O
(b) Reflect design in M1
(c) Reflect design in M2

What progress have you made? (p 353)

1

2 (a) A translation 3 units right, 3 down
(b) A reflection in M1

Practice booklet

Sections A, B and C (p 111)

1. (a) C (b) E
   (c) Reflection in line M_4
   (d) A translations 2 units to the right and 2 units down
   (e) A translation of 3 units to the left

2. (a), (b), (c)

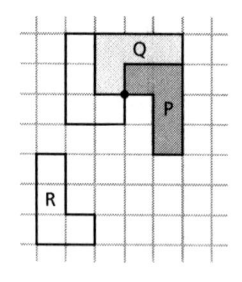

3. (a), (b), (c), (d)

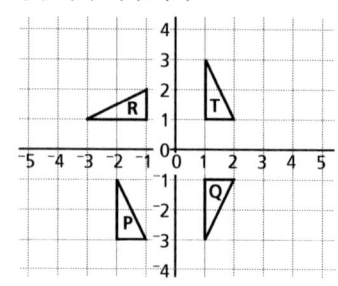

4. 1 unit left, 2 units down

48 Negative numbers 7C/18

> **Practice booklet** pages 112 and 113

A The joke contest – adding (p 354)

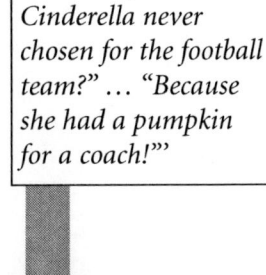

> *'The best joke we had was: "Why was Cinderella never chosen for the football team?" ... "Because she had a pumpkin for a coach!"'*

◊ It is important to make the distinction between ⁻3 (a quantity which in itself is negative, such as a temperature or a negative score) and – 3 (which means subtract the positive number 3).

◊ When trialling this unit, several teachers held joke contests, with pupils as judges awarding points from ⁻10 to 10. Teachers who did this found their pupils then understood adding and subtracting directed numbers better. If you need to tell some jokes to set the ball rolling, here are some suggestions (of suitably varied quality).

Q: What do you call a camel with three humps?   A: Humphrey
Q: What did the policeman say to his tummy?   A: You're under a vest.
Q: What did the Spanish farmer say to his chickens?   A: Olé!

Focus first on the fact that when you add a set of numbers together any negative number in the set pulls the score down. Check that pupils realise that, for example, ⁻3 + 5 is equal to 5 + ⁻3.

These are the totals for the pictures on the lower half of page 354.

A 2   B ⁻8   C ⁻2   D 1   E ⁻2

◊ It may be tempting to treat (for example) 5 + ⁻1 as 5 – 1. They give the same answer but they are different things: the first is a preparation for substituting (say) $a = 5$ and $b = ⁻1$ into the formula $p = a + b$; the second is not.

A5, A6 If pupils have not met magic squares before, you may need to demonstrate with one, for example, that has the numbers 1 to 9 in it; its magic number will be 15.

B The joke contest – subtracting (p 356)

◊ Again, the idea of taking off a negative score is best understood if it is acted out by pupils. You can tell some story about judges being discovered taking bribes to justify the removal of their scores.

48 Negative numbers • 259

A The joke contest – adding (p 354)

A1 (a) ⁻1 (b) ⁻11 (c) 0 (d) ⁻1

A2 (a) 3 (b) 1 (c) 6
(d) ⁻4 (e) ⁻3 (f) ⁻3
(g) 6 (h) 9 (i) 3

A3 3°C

A4 ⁻3°C

A5 (a)
4	⁻1	0
⁻3	1	5
2	3	⁻2

(b)
1	6	⁻1
0	2	4
5	⁻2	3

(c)
⁻2	3	⁻4
⁻3	⁻1	1
2	⁻5	0

*A6
1	⁻6	⁻1
⁻4	⁻2	0
⁻3	2	⁻5

or a square that is a rotation or reflection of this.

Note that the square above can be produced by subtracting 7 from each entry in a 1 to 9 square:

8	1	6
3	5	7
4	9	2

B The joke contest – subtracting (p 356)

B1 1 – ⁻5 = 6

B2 10 – 12 = ⁻2

B3 3 – ⁻4 = 7

B4 (a) 2 (b) 4 (c) 2 – ⁻2 = 4

B5 (a) ⁻9 (b) ⁻1 (c) ⁻9 – ⁻8 = ⁻1

B6 Make the number positive then add it.

B7 (a) 12 (b) 22 (c) 32
(d) 60 (e) 40

B8 (a) 6 (b) ⁻6 (c) ⁻2
(d) 25 (e) ⁻17

B9 (a) 4 (b) ⁻6 (c) 2
(d) ⁻15 (e) ⁻9

B10 ⁻13

B11 ⁻10 – ⁻8 = ⁻2

B12 (a) 6 (b) ⁻16 (c) 20
(d) 269 (e) 70 (f) 833

B13 (a) ⁻13 (b) 20 (c) 9 (d) 89

B14 (a) 110 – ⁻155 (b) 265 degrees

B15 (a) 50 (b) ⁻40 (c) 15
(d) ⁻23 (e) ⁻17 (f) ⁻3
(g) ⁻18 (h) ⁻5 (i) ⁻3

What progress have you made? (p 358)

1 (a) 2 (b) ⁻11 (c) 0 (d) ⁻2

2
1	⁻9	5
3	⁻1	⁻5
⁻7	7	⁻3

3 (a) 10 (b) ⁻10 (c) 4
(d) 0 (e) ⁻11 (f) ⁻9

4 (a) ⁻23 (b) 18

5 (a) ⁻11 + ⁻9 = ⁻20
(b) 8 – ⁻4 = 12
(c) ⁻4 – ⁻1 = ⁻3
(d) ⁻8 – ⁻12 = 4

Practice booklet

Sections A and B (p 112)

1 (a) ⁻1 (b) 0 (c) ⁻4
(d) ⁻6 (e) 7 (f) 3

2 (a)
-3	2	-5
-4	-2	0
1	-6	-1

(b)
0	5	-2
-1	1	3
4	-3	2

(c)
3	-2	-1
-4	0	4
1	2	-3

3 (a) 0 (b) ⁻6 (c) ⁻2 (d) 1

4 13

5 (a) 14 (b) 19 (c) 34 (d) 28
  (e) 7   (f) ⁻15 (g) ⁻5 (h) ⁻2
  (i) 25 (j) 6   (k) ⁻6 (l) ⁻10

6 (a) 7   (b) 10 (c) 0   (d) 246

7 (a) 166 (b) 19 (c) 21 (d) ⁻15

49 Oral questions: money 2 (p 359)   7T/38

> 'It was the last 15 minutes of the lesson and lunch time was approaching. Most pupils did very well making only one or two small mistakes or getting all correct. It was very good diagnostically.'

◊ The information on the pupil's page is intended for use as a context for orally given questions. Some sample questions are given after these notes.

◊ An initial discussion of pupils' own calculating methods may be beneficial, especially ways of calculating mentally with money.

◊ As a follow-up activity, pupils can invent their own questions for the rest of the class. In one school, each pupil wrote their own question on a slip of paper and put the answer on the back. The questions were collected together and read out for the whole class to answer.

1	What is the total cost of a note pad and an eraser?	£0.72 or 72p
2	Find the cost of four packs of ink cartridges.	£4.28
3	Find the total cost of the two most expensive items.	£16.30
4	How much is a ruler and a note pad?	£1.36
5	Find the cost of four note pads.	£2.28
6	Find the cost of ten rulers.	£7.90
7	I buy the sticky tape and the scissors. How much is this?	£4.34
8	The shop has a half-price sale. What is the sale price of the hole punch?	£4.25
9	What is the cost of two rulers?	£1.58
10	I buy a hole punch and a pencil case. How much change will I get from £10?	£0.40 or 40p
11	How much would the stapler cost if the price was reduced by 50p?	£7.30
12	What is the cost of two geometry sets?	£10.66
13	What is the difference in cost between the dearest and cheapest article?	£8.35
14	How much will pencils for a class of 30 cost?	£3.35
15	How many erasers could you buy for £1?	6
16	How much is a geometry set and a ruler?	£6.12
17	How much for the stapler and colouring pencils?	£11.00
18	One Stick-it note has a width of 4 cm and an area of 20 cm^2. What is the length of the note?	5 cm
19	How many rulers can you buy for £3.00?	3

20	What is the total cost of the geometry set, the hole punch and the stapler?	£21.63
21	How much do you think one pencil should be sold for?	11p or 12p
22	The shop has a half-price sale. What is the sale price of the colouring pencils?	£1.60
23	There are five maths teachers in a school. How much would it cost to buy each teacher a hole punch?	£42.50
24	How many packs of ink cartridges could you buy with £12 and how much money would you have left?	11, 23p left

50 Big numbers

A Lots of dots! (p 360)

B Time (p 361)

C Money (p 361)

These questions can be tackled without a calculator, as exercises in rough estimation. Pupils may need to measure a telephone box and do some research to find the distance from Land's End to John o'Groats.

A Lots of dots! (p 360)

There are 100 groups of 100 dots, or 10 000 dots on a page.
The book has 100 pages.

B Time (p 361)

B1 A million hours is just over 114 years. Few people have lived that long.

B2 A billion seconds is about 31.7 years from now.

B3 A billion minutes is about 1901 years ago if you take a year to be $365\frac{1}{4}$ days (1903 years if you use 365 days).

B4 A million seconds is about $11\frac{1}{2}$ days, so you would not have had a birthday yet.

B5 If it takes a second on average to count each number, then it takes about $11\frac{1}{2}$ days.

B6 The time line of the universe would be 15 km long.
The part representing human existence would be 2 metres long.

C Money (p 361)

C1 A pound coin's diameter is about 2.2 cm. A million of them would be 22 km, or about 13.75 miles long.
It would take over $4\frac{1}{2}$ hours to walk past.

C2 The distance is about 850 miles, or 1350 km, by road.
Dividing by 22 km for one million coins gives over 61 million coins.

C3 A pile of ten £1 coins is about 3.3 cm tall. A million would be 3.3 km tall, much taller than the Eiffel Tower (300 m).

C4 A phone box is about 208 cm high, so each pile would have about 630 coins. The box is about 75 cm by 70 cm at the base, so you could fit about 34 piles by 31 piles on a square grid, making a total of about 664 000 coins. By putting the coins on a triangular grid you could fit more in, but not enough to make a million.

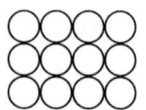

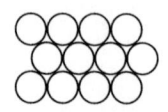

C5 Ten £1 coins weigh 90 g. If your weight is, for example, 45 kg, then you would be worth £5000.

51 Functions and graphs

7C/41

Essential	**Optional**
2 mm graph paper	OHP
Squared paper	

Practice booklet pages 114 and 115 (graph paper needed)

A From table to graph (p 362)

2 mm graph paper

◊ The main points to be covered in the introduction are:
- We can make a table from a rule given in words.
- The values in the table can be plotted as points.
- When the variables are continuous, a line can be drawn through the points to show the relationship between the variables.

You should point out that real-life situations rarely give precise straight-line graphs (except in cases where linearity is built in, as for example with currency conversion graphs).

B Using formulas (p 364)

B5 For the first time in this unit the pupil has to find a rule from some number pairs.

C Functions (p 366)

Optional: OHP

The activity referred to on page 367 is similar to unit 27 'Spot the rule'. You will need a large grid on the board or OHP with x- and y-axes.

◊ You can start the activity by telling the class that you are thinking of a rule linking x and y, for example $y = x + 2$. Ask a member of the class to give you a value of x; work out y and plot the corresponding point on the grid. Continue until someone can tell you your rule.

Then ask a pupil to take over with a rule of their own.

◊ If pupils are adventurous and use rules involving, for example, squaring, then you can ask what kinds of rule give points which lie in a straight line.

◊ If nobody else does so, use a rule like $x + y = 6$.

51 Functions and graphs • 265

◊ Ask pupils if they see any similarities and differences between earlier work on sequences and this work. They should be able to appreciate that
- in the case of a sequence, n is restricted to being a whole number (discrete), whereas x is continuous
- the values of y for $x = 1, 2, 3, 4, \ldots$ form a sequence (which helps when pupils are trying to find the equation of a linear graph).

D Negative numbers on graphs (p 368)

> Squared paper

◊ This work shows that calculations involving negative numbers continue a straight line in the expected way.

A From table to graph (p 362)

A1 (a)

Gas in tank (kg)	Hours away from base
1	1
2	3
3	5
4	7
5	9
6	11
7	13

(b) The pupil's graph of points from the table, joined with a straight line

(c) 8 hours   (d) $5\frac{1}{2}$ kg

A2 (a)

t	0	1	2	3	4	5	6
h	20	50	80	110	140	170	200

(b) The pupil's graph of points from the table

(c) 65 cm

(d) $\frac{1}{2}$ hour

(e) 2 hours 40 minutes

(f) 95

A3 (a)

t	0	1	2	3	4	5	6
l	28	26	24	22	20	18	16

(b) The pupil's graph of points from the table

(c) 21 cm   (d) $5\frac{1}{2}$ hours

(e) $1\frac{1}{2}$ (f) 19

B Using formulas (p 364)

B1 (a) $c = 4d + 2$ (b) $c = 3d + 3$
(c) $c = 5d$

B2 (a) £11   (b) £23   (c) £51
(d) £71   (e) £13

B3 (a)

d	1	2	3	4	5	6	7
c	6	11	16	21	26	31	36

(b) The pupil's graph of points from the table, labelled $c = 5d + 1$

(c) £28.50

(d) $3\frac{1}{2}$ miles

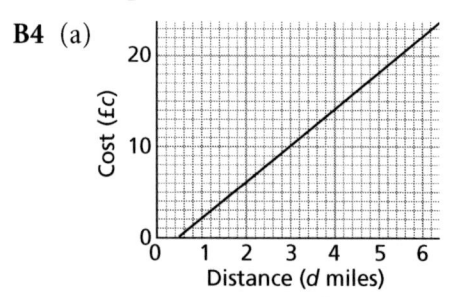

B4 (a) The pupil's graph showing Cost (£c) vs Distance (d miles)

(b) (i) 5 miles (ii) $3\frac{1}{2}$ miles
(iii) $4\frac{1}{4}$ miles

(c) Nothing!

(d) Applying the rule strictly you would pay ⁻£1, meaning perhaps that the taxi firm would pay you £1; but this is not very likely!

B5 (a)

Weight on spring, w kg	0	1	2	3	4	5
Length of spring, l cm	15	25	35	45	55	65

(b) $l = 10w + 15$

(c)

Weight on spring, w kg	0	1	2	3	4	5
Length of spring, w cm	20	25	30	35	40	45

(d) $l = 5w + 20$

C Functions (p 366)

C1 (a)

x	0	1	2	3	4	5	6
y	1	2	3	4	5	6	7

(b)

x	1	2	3	4
y	1	3	5	7

(c)

x	0	1	2	3	4	5
y	5	4	3	2	1	0

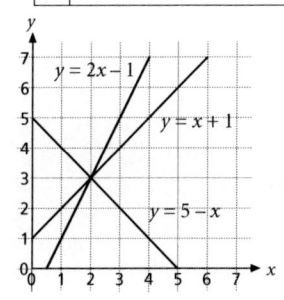

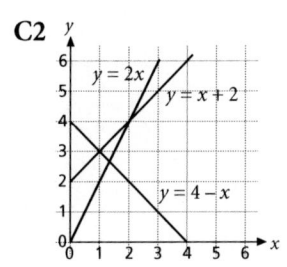

C2

C3

x	0	1	2	3
y	4	7	10	13

$y = 3x + 4$

C4

x	0	1	2	3	4
y	10	9	8	7	6

This was Ted's rule.

C5 $y = 2x + 1$

C6 $y = x + 3$

C7 Owen

C8 Trevor

D Negative numbers from graphs (p 368)

D1 (a)

x	0	1	2	3	4	5
y	⁻7	⁻5	⁻3	⁻1	1	3

(b) The pupil's graph of points from the table

(c) 2

(d) $2\frac{1}{2}$

D2 (a)

x	0	1	2	3	4	5
y	⁻3	⁻2	⁻1	0	1	2

(b) The pupil's graph of points from the table

D3 (a)

x	⁻3	⁻2	⁻1	0	1	2	3
y	⁻2	⁻1	0	1	2	3	4

(b) The pupil's graph of points from the table

D4 (a)

x	⁻3	⁻2	⁻1	0	1	2	3
y	⁻4	⁻3	⁻2	⁻1	0	1	2

(b) The pupil's graph of points from the table

What progress have you made? (p 369)

1 (a)

t	0	1	2	3	4	5
c	10	25	40	55	70	85

(b) The pupil's graph of points from the table

2 $3\frac{1}{2}$ hours

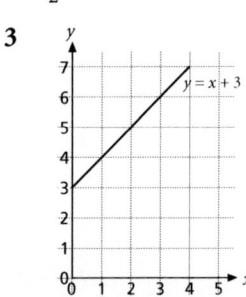

3

51 Functions and graphs • 267

4

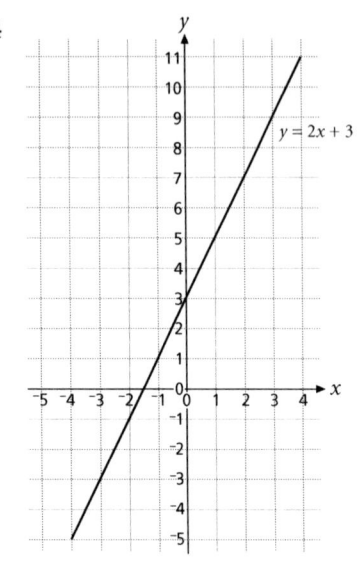

[Graph showing line $y = 2x + 3$]

5 (a) $y = 2x + 2$ (b) $y = 7 - x$

Practice booklet

Sections A and B (p 114)

1 (a)
t	0	1	2	3	4	5
d	80	100	**120**	**140**	**160**	**180**

(b) $d = 80 + 20t$

(c) The pupil's graph of points from the table

(d) $t = 2.4$

2 (a)
t	0	1	2	3	4	5
d	30	**26**	**22**	**18**	**14**	**10**

(b) The pupil's graph of points from the table

(c) 3.5 minutes

Sections C and D (p 115)

1 (a)
x	0	1	2	3	4	5	6	7	8
y	8	7	6	5	4	3	2	1	0

(b) The pupil's labelled graph of $y = 8 - x$

2 (a)
x	⁻4	⁻3	⁻2	⁻1	0	1	2	3	4
y	⁻12	⁻10	⁻8	⁻6	⁻4	⁻2	0	2	4

(b) The pupil's labelled graph of $y = 2x - 4$

3 (a) $y = 3x + 2$ (b) $y = 7$
  (c) $y = x - 2$ (d) $y = 6 - x$

4 $h = 35 - 5w$

268 • 51 Functions and graphs

52 Multiples and factors

7C/13

Essential	Optional
Sheets 134 and 174 to 176 Dice, counters (two colours) Sheet 177 (or other polygon tiles)	OHP transparency of sheet 175

Practice booklet pages 116 to 118

A Multiples (p 370)

Nasty multiples

'This game is very good. I enjoyed playing it and the pupils sometimes beat me.'

Each pair needs sheet 174, a dice and several counters of two colours

After revising the idea of a multiple, pupils can consolidate their knowledge by playing the game.

B Factors (p 371)

Sheets 175 and 134 (for the sieve of Eratosthenes)
Optional: OHP transparency of sheet 175

The factor chart makes it possible to see and discuss patterns.

◊ You could get the class to make a large factor chart on squared paper for numbers up to, say, 80. Groups could be given different ranges of numbers to work on.

One teacher tried this and reported:

'I split the class to work in fours, one pair to do factors of 1–45, the other 40–80. I was then going to overlap them for display. Students found the activity difficult, some didn't immediately see the patterns. 40–80 was a disaster area! For 1–45, about half the pairs managed successfully. BUT ... the quality of discussion and language and understanding the activity generated was EXCELLENT, the work for display not very good!'

◊ A popular game (not referred to in the pupil's book) is 'Factor bingo'.

Ask each pupil to write down seven numbers which are to be factors. They can repeat (for example, 2, 2, 3, 3, 4, 6, 7). They are not allowed to use 1, but if they suggest it, reward their perspicacity!

Make a list of 'multiples' for yourself and read them out one at a time. Include the occasional prime number!

52 Multiples and factors • 269

Pupils can cross out any factor of the number you read out. They are not allowed to cross out a number more than once. For example, if a pupil's list includes 2, 2, 4, 8 and the number read out is 16, they can cross out one of the 2s, the 4 and the 8.

For the first couple of games you can check after each number what factors could have been crossed off.

C **Multiples and factors** (p 372)

> Sheet 176 (for the 'Multiples and factors maze')

D **Divisibility** (p 373)

> Sheet 134 (for question D5)

E **Common multiples** (p 374)

F **Common factors** (p 375)

G **Common factors and multiples** (p 376)

H **Polygon wheels** (p 377)

> Polygon tiles, from sheet 177 or elsewhere (for example ATM polygon mats); an OHP is very useful.

◊ If you demonstrate with polygons on the OHP, cut a small slit in an edge to represent the dot.

After a couple of turns of the square about the pentagon, you could ask:
- Will the dots ever come together again?
- If so, how many turns do you think it will take, and why?

◊ Pupils may jump to the conclusion that the number of turns is found by multiplying the numbers of sides of the two polygons. If so, you can ask them to find out if this is always true.

The number of turns is the lowest common multiple of both numbers of sides. For example, for a square rolling round a hexagon the number of turns is the lowest common multiple of 4 and 6, that is 12.

A Multiples (p 370)

A1 3, 9, 12, 30, 33

A2 8, 16, 24, 48, 56

A3 (a) Any five multiples of 10
(b) 70, 90, 100, 110, 170, 220
(c) The last digit is 0.

A4 (a) 6, 8, 18, 20, 24, 36, 60
(b) 6, 15, 18, 24, 36, 60
(c) 8, 20, 24, 36, 60
(d) 15, 20, 60

A5 33p, 59p, 66p

A6 (a) 12 is a multiple of 1, 2, 3, 4, 6 and 12.
(b) 15 is a multiple of 1, 3, 5 and 15.
(c) 24 is a multiple of 1, 2, 3, 4, 6, 8, 12 and 24.
(d) 30 is a multiple of 1, 2, 3, 5, 6, 10, 15 and 30.

B Factors (p 371)

B1 (a) 1 2 3 4 6 12
(b) 1 2 3 6 9 18
(c) 1 2 3 4 6 8 12 24
(d) 1 2 3 5 6 10 15 30

B2 (a) 1 2 3 4 6 9 12 18 36

One factor pairs off with itself.
(b) 36 has an **odd** number of factors.
(c) Any three square numbers
(d) They are square numbers, with one factor pairing off with itself.

B3 1 7

There are only two factors, 1 and the number itself.

B4 2, 3, 5, 7, 11, 13

B5 23 and 29

B6 Possible numbers are 13, 17, 37 and so on.

C Multiples and factors (p 372)

C1 (a) 6 is a **multiple** of 2.
(b) 6 is a **factor** of 30.
(c) 10 is a **factor** of 60.
(d) 8 is a **factor** of 32.
(e) 8 is a **multiple** of 4.
(f) 36 is a **multiple** of 9.

C2 (a) 1, 5, 25
(b) 25, 50, 75, 100, 150

C3 40, 80, 120

C4 1, 2, 3, 4, 5, 6, 10, 12, 60

C5 (a) The pupil's four multiples of 6
(b) 1, 2, 3, 6

Multiples and factors maze

36	10	4	5←35	10	12	5
18	20	16	28	20	7	50	27
50	21	35	6→48	42←6	16
18	4	30	21	9	8	54	40
32	9	16→2	15	56	12	9
8←24	15	9	27	7	18	4
14	10	6	20	28	20	30→6
3 →12	8	16	6	4	10	8

D Divisibility (p 373)

D1 (a) 562, 3334, 3108, 874, 678, 400 098, 67 924
(b) If the last digit is 0, 2, 4, 6 or 8, the number is divisible by 2.
(c) The last digit is 6, so the number is divisible by 2.

D2 (a) If the last digit is 0, then the number is divisible by 10.
(b) If the last digit is 0 or 5, then the number is divisible by 5.
(c) (i) 2390, 60, 1200, 7810
(ii) 2390, 765, 7415, 60, 7810, 1200

D3 (a) 782 654 ÷ 3 = 260 884.666 66…
The decimal part tells us that
782 654 is not divisible by 3.

(b) (i)

4056	Yes	15	Yes
1101	Yes	3	Yes
692	No	17	No
9218	No	20	No

(ii) The pupil's numbers
in the table

(c) If the sum of the digits is divisible by
3, then the number is divisible by 3.

(d) 3108, 678, 4893, 400 098

D4 (a) The pupil's numbers that are
divisible by 9 and their digit sums

(b) The pupil's numbers that are not
divisible by 9 and their digit sums

(c) If the digit sum is divisible by 9, then
the number is divisible by 9.

(d) 504, 21 348, 39 285, 61 236

D5 (a) The pupil's hundred square with
each multiple of 2 underlined

(b) The pupil's hundred square with a
circle round each multiple of 3

(c) Numbers divisible by 6 are divisible
by 2 and 3.

(d) (i) 106, 138, 1314, 5012, 135 912
(ii) 138, 1314, 213, 135 912
(iii) 138, 1314, 135 912

*D6 (a) If the last two digits are divisible by
4, then the number is divisible by 4.

(b) If the last three digits are divisible by
8, then the number is divisible by 8.

E Common multiples (p 374)

E1 (a) Multiples of 3: 3, 6, 9, 12, 15, 18,
21, 24, 27, 30
Multiples of 5: 5, 10, 15, 20, 25,
30, 35, 40, 45, 50

(b) Three common multiples of 3 and 5
such as 15, 30 and 45

(c) 15

E2 (a) 4, 8, 12, 16, 20, 24, 28, 32, 36, 40
(b) 6, 12, 18, 24, 30, 36, 42, 48, 54, 60
(c) Three common multiples of 4 and 6
such as 12, 24 and 36
(d) 12

E3 (a) Four common multiples of 2 and 5
such as 10, 20, 30 and 40
(b) 10

E4 (a) 6   (b) 18   (c) 6

E5 (a) 30   (b) 12   (c) 6

F Common factors (p 375)

F1 (a) 1, 3, 5, 15   (b) 1, 2, 4, 5, 10, 20
(c) 1 and 5   (d) 5

F2 (a) 1, 2, 3 and 6 (b) 6

F3 2

F4 (a) 3   (b) 8   (c) 1

F5 4

G Common factors and multiples (p 376)

G1 (a) 20   (b) 2

G2 (a) 24   (b) 1

G3 20 seconds

G4 4p

G5 30 sweets

G6 6 cm by 6 cm square tiles

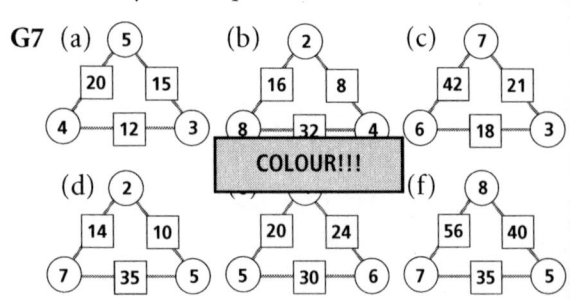

G7 (a)–(f) COLOUR!!!

What progress have you made? (p 377)

1 The pupil's three multiples of 5

2 Three factors of 24 from 1, 2, 3, 4, 6, 8, 12 and 24

3 2, 3, 5, 7, 11, 13, 17 and 19

4 The last digit (6) tells you that the number is even and so it cannot be prime.

5 21 6 12

Practice booklet

Sections A, B and C (p 116)

1 (a) 6, 8, 10, 30, 90, 100, 110
 (b) 10, 30, 75, 90, 100, 110
 (c) 6, 30, 90
 (d) 10, 30, 90, 100, 110

2 (a) 1, 2, 4, 8   (b) 1, 3, 9
 (c) 1, 3, 5, 15
 (d) 1, 2, 4, 5, 10, 20
 (e) 1, 3

3 33, 63

4 (a) 5 is a factor of 10.
 (b) 20 is a multiple of 4.
 (c) 18 is a multiple of 6.
 (d) 18 is a multiple of 3.
 (e) 9 is a factor of 18.
 (f) 3 is a factor of 6.

5 1 2 **4** 5 **10** 20 25 **50 100**

6 6, 36

7 1, 2, 3, 6, 7, 14, 21, 42

8 2, 7, 11, 13, 19, 23, 29

Section D (p 117)

1 (a) 340, 156, 120, 7000, 902, 650, 4008
 (b) 340, 265, 6755, 120, 7000, 3265, 650
 (c) 340, 120, 7000, 650

2 (a)

Number	Digit sum
471	12
1260	9
856	19
4734	18
1382	14
5853	21
109 832	23

 (b) 471, 1260, 4734, 5853
 (c) 1260, 4734

3 (a) 402, 347 622, 57 674
 (b) 402, 347 622, 5013, 9831
 (c) 402, 347 622

4 561

5 487

Sections E, F and G (p 117)

1 (a) 4, 8, 12, 16, 20, 24, 28, 32, 36, 40
 (b) 3, 6, 9, 12, 15, 18, 21, 24, 27, 30
 (c) Any three common multiples of 4 and 3 such as 12, 24 and 36
 (d) 12

2 (a) Any four common multiples of 2 and 6 such as 6, 12, 18 and 24
 (b) 6

3 (a) 30   (b) 14   (c) 8

4 (a) 1, 2, 7, 14   (b) 1, 5, 7, 35
 (c) 1, 7   (d) 7

5 (a) 1, 2, 4   (b) 4

6 (a) 5   (b) 4   (c) 1

7 40 seconds

8 All possible prices are 1p, 2p, 3p, 4p, 5p, 6p, 10p, 12p, 15p, 20p, 30p and 60p

9 28 is the smallest possible number of pupils. As 56 (the next possible number) is rather large for a class, the number of pupils is probably 28.

*10 1

53 Know your calculator

7C/36

p 378	**A** In order	The priority rules for multiplication, division, addition and subtraction
p 379	**B** Brackets	
p 380	**C** A thin dividing line	Interpreting expressions such as $\frac{4}{2}$ and $\frac{5+1}{3}$.
p 381	**D** All keyed up	Using the brackets keys and the '=' key in the middle of a calculation
p 383	**E** Memory	Using a calculator's memory facilities
p 383	**F** Squares	Using the square key
p 384	**G** Square roots	Square roots including use of the square root key
p 385	**H** Negative numbers	Using the 'sign-change' key to work with negative numbers
p 386	**I** Complex calculations	Calculations that mix the four operations, negative numbers, squares and square roots

> **Essential**
> Scientific calculators
> Sheets 179 to 181 (for the game 'Operation 3062')
> **Practice booklet** pages 119 to 122

A In order (p 378)

Pupils investigate the rules used by a scientific calculator to evaluate expressions that use two operations to arrive at the priority rules for dealing with multiplication, division, addition and subtraction.

The game 'Operation 3062' consolidates the use of these rules.

> Scientific calculators (one for each pupil)

◊ You could begin by asking pupils to predict what they think their calculators will give for each set of key presses on page 378. Pupils find the result of each set of key presses (remind them they need to press the '=' key or 'ENTER' key at the end) and try to describe the rules they think the calculator uses to evaluate expressions that use any combination of

274 • 53 *Know your calculator*

the four operations. They should try their rules out on their own expressions. Working in groups, each group could try to produce a clear statement of the rules they think the calculator uses.

A brief statement of the rules could be

- You multiply or divide before you add or subtract.
- Otherwise, work from left to right.

Once pupils understand these rules, point out that they are widely used in numerical calculations and they should use them for the rest of the unit.

◊ Some calculators use the symbols * and / for × and ÷ respectively. Pupils with these calculators need to interpret the × key press as * and the ÷ key press as /.

A6 Pupils could consider how many different results are possible. They could make up their own puzzles like this for someone else to solve.

A7 Operation 3062

This game consolidates the priority rules for multiplication, division, addition and subtraction.

> Sheet 179 (one copy of the rules and board for each group)
> Sheet 180 (two sheets on card for each group)
> Sheet 181 (one sheet on card for each group)

◊ Calculators should not be used for the game.

◊ Sheets 180 and 181 need to be copied on to card so that the operations and numbers do not show through.

◊ Here is full set of results for 'Operation 3062':

$30 + 6 + 2 = 38$	$30 - 6 + 2 = 26$	$30 \times 6 + 2 = 182$	$30 \div 6 + 2 = 7$
$30 + 6 - 2 = 34$	$30 - 6 - 2 = 22$	$30 \times 6 - 2 = 178$	$30 \div 6 - 2 = 3$
$30 + 6 \times 2 = 42$	$30 - 6 \times 2 = 18$	$30 \times 6 \times 2 = 360$	$30 \div 6 \times 2 = 10$
$30 + 6 \div 2 = 33$	$30 - 6 \div 2 = 27$	$30 \times 6 \div 2 = 90$	$30 \div 6 \div 2 = 2.5$

◊ You may wish to adapt the board and result cards so that pupils can play the game with a different set of numbers (that could lead to negative results).

One suggestion is: | 16 | | 8 | | 4 | = |

with result cards: 28, 20, 48, 18, 4, 12
⁻16, 14, 512, 132, 124, 32
0.5, 8, 6, ⁻2

◊ Pupils could design game boards and result cards for other groups to use.

B Brackets (p 379)

Pupils do the questions without a calculator but could use a calculator to check their results.

C A thin dividing line (p 380)

Some calculators use a line to show division. Pupils with calculators like this should not use them for C1 and C2.

D All keyed up (p 381)

◊ Not all brackets keys are the same. For example on some calculators the brackets keys look like this: [(---] [---)]

Make sure pupils know which keys to use on their calculators.

D5 Pupils could write an appropriate expression for each set of key presses.

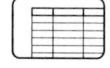

◊ Pupils could use a spreadsheet to investigate the value of the expression $a \Diamond b \Diamond c$ for various values of a, b and c where each diamond can be replaced by any one of the four operations and brackets can be used. The spreadsheet could be set up as shown below.

	A	B	C
1	10		
2	5		
3	2		
4			
5	=A1+A2+A3	=(A1+A2)+A3	=A1+(A2+A3)
6	=A1+A2−A3	=(A1+A2)−A3	=A1+(A2−A3)
7	=A1+A2*A3	=(A1+A2)*A3	=A1+(A2*A3)
8	=A1+A2/A3	=(A1+A2)/A3	=A1+(A2/A3)
9			
10	=A1−A2+A3	=(A1−A2)+A3	=A1−(A2+A3)
11	=A1−A2−A3	=(A1−A2)−A3	=A1−(A2−A3)
12	=A1−A2*A3	=(A1−A2)*A3	=A1−(A2*A3)
13	=A1−A2/A3	=(A1−A2)/A3	
14			
15	=A1*A2+A3	=(A1*A2)+A3	
16	=A1*A2−A3		
17	=A1*A2		
18	=A1*A		

The investigation will be much easier if pupils have a separate note of the expressions in each cell possibly in the form below:

	A	B	C
1	a		
2	b		
3	c		
4			
5	a + b + c	(a + b) + c	a + (b + c)
6	a + b − c	(a + b) − c	a + (b − c)
7	a + b × c	(a + b) × c	a + (b × c)
8	a + b ÷ c	(a + b) ÷ c	a + (b ÷ c)
9			
10	a − b + c	(a − b) + c	a − (b + c)
11	a − b − c	(a − b) − c	a − (b − c)
12	a − b × c	(a − b) × c	(b × c)
13	a − b ÷ c	(a − b) ÷ c	
14			
15	a × b + c	(a − b) +	
16	a × b − c		
17	a × b		
18	a × b		

One outcome is confirmation of the priority rules,
for example that $a + b \times c = a + (b \times c)$ for all values of a, b and c.

Pupils could consider:
- Which values for a, b and c give a set of positive results?
- Which values give a set of integer results?
- Which values give a set of non-recurring decimals?
- What happens if a, b or c has a value of 1? or 0?
- Which values give 16 different answers in the first column?
  Which values give repeats and why?
- What happens if two values are equal? What about three equal values?
- Which expressions always have the same value and why?
  For example, $a - b - c = a - (b + c)$ and
  $a \div b \div c = a \div (b \times c)$
- Why do some rows always give three identical results?
  For example, $a + b + c = (a + b) + c = a + (b + c)$
  $a \times b \div c = (a \times b) \div c = a \times (b \div c)$

E **Memory** (p 383)

◊ Some calculators use the same key to store and recall so you may have to discuss the 'shift' or '2nd function' key.

53 Know your calculator • 277

E2 Emphasise that pupils can use any method they feel confident with: for example, some may feel much happier using the brackets keys throughout. More confident pupils could think about different ways to carry out each calculation and hence check their results.

F Squares (p 383)

Some calculators use the same key for squares and square roots so you may have to discuss the 'shift' or '2nd function' key.

G Square roots (p 384)

H Negative numbers (p 385)

I Complex calculations (p 386)

A In order (p 378)

A1 (a) 5 (b) 7 (c) 7* (d) 7*
(e) 0 (f) 6

* Pupils who give 8 and 5 as their answers for (c) and (d) respectively are probably consistently working from left to right.

A2 A, B, E and F

A3 The pupil's sets of key presses with result 8

A4 (a) 4 (b) 7 (c) 10 (d) 2
(e) 4 (f) 5 (g) 8 (h) 30
(i) 23 (j) 12 (k) 9 (l) 2
(m) 20 (n) 5 (o) 13

A5 (a) 6 (b) 6 (c) 2 (d) 4
(e) 9 (f) 8 (g) 12 (h) 4
(i) 5

A6 (a) 12 + 6 − 2 (b) 12 × 6 − 2
(c) 12 + 6 × 2 (d) 12 + 6 ÷ 2

B Brackets (p 379)

B1 (a) 12 (b) 2 (c) 2 (d) 15
(e) 2 (f) 2 (g) 4 (h) 16
(i) 3

B2 (a) 12 (b) 14 (c) 5 (d) 11
(e) 5 (f) 3 (g) 5 (h) 7

B3 (a) 3 (b) 6 (c) 3 (d) 11

B4 A, C and E

C A thin dividing line (p 380)

C1 A and C   B and G   D and F   H and I

C2 (a) $10 + \frac{6}{2}$ (b) $\frac{18 - 2}{4}$
(c) $\frac{8 + 4}{3}$ (d) $12 - \frac{10}{5}$
(e) $\frac{5}{3 - 1}$ (f) $\frac{12}{4} + 2$

C3 (a) 7 (b) 14 (c) 4 (d) 3
(e) 2 (f) 12 (g) 10 (h) 2
(i) 15

278 • 53 Know your calculator

D All keyed up (p 381)

D1 (a) 2 (b) 15 (c) 5

D2 (a) 5* (b) 5 (c) 37
(d) 4 (e) 35 (f) 5

* Pupils who give 19 as their answer for (a) have probably keyed in 16 + 24 ÷ 8 omitting the necessary brackets.

D3 (a) 9 (b) 13 (c) 9
(d) 4 (e) 28 (f) 1.6

D4 (a) (i) 19 (ii) 27 (iii) 9 (iv) 5
(b) The pupil's description

D5 (a) 15 (b) 5 (c) 2

D6 A and C

D7 [9][−][3][=][÷][2]
or
[(][9][−][3][)][÷][2]

D8 (a) 9 (b) 46 (c) 2.5
(d) 22.1 (e) 3.8 (f) 0.5
(g) 7 (h) 4 (i) 15

E Memory (p 383)

E1 (a) 59 (b) 2 (c) 0.25
(d) 13.5 (e) 8.2 (f) 100

E2 (a) 1.2 (b) 25.97 (c) 60
(d) 6.5 (e) 469 (f) 42

E3 (a) 1.5 gallons (b) 5.3 gallons
(c) 2.7 gallons (d) 39.6 gallons

F Squares (p 383)

F1 (a) 441 (b) 3969 (c) 11664
(d) 96721 (e) 2500

F2 (a) 19 (b) 28 (c) 300 (d) 67

F3 625

F4 (a) 2304 (48^2) (b) 2116 (46^2)

F5 (a) 818 m^2 (b) 456 m^2

F6 Yes, 17 stones on each side

F7 No

F8 (a) 10.24 (b) 32.49 (c) 2.1316
(d) 0.64 (e) 0.0144

F9 (a) 1.9 (b) 4.8 (c) 0.9 (d) 1.05

G Square roots (p 384)

G1 (a) 5 (b) 2 (c) 7
(d) 10 (e) 1

G2 (a) 4 (b) 9 (c) 6
(d) 8 (e) 12

G3 (a) 11 (b) 15 (c) 20
(d) 69 (e) 59

G4 (a) 6.5 (b) 3.7 (c) 1.28
(d) 0.6 (e) 0.04

G5 (a) 3000
(b) The pupil's own estimates; if each person stands in a metre square the board will be 3 km wide.

H Negative numbers (p 385)

H1 (a) (i) 5 (ii) 4 (iii) ⁻5 (iv) ⁻6
(b) The pupil's checks

H2 (a) 1 (b) ⁻5 (c) ⁻8
(d) ⁻4 (e) ⁻6 (f) ⁻7
(g) ⁻4 (h) ⁻20

H3 (a) 10 (b) 8 (c) ⁻3
(d) 1 (e) ⁻8.5 (f) ⁻8.1
(g) 4.9 (h) ⁻4.1

H4 (a) 2 (b) ⁻2 (c) 5
(d) ⁻7 (e) ⁻5 (f) ⁻5

I Complex calculations (p 386)

I1 (a) 25 (b) 4 (c) 50 (d) 22
(e) 3 (f) 25 (g) 35 (h) 35
(i) 12 (j) 11 (k) 4 (l) 3

I2 (a) 18 (b) 36 (c) 3 (d) 81

13 (a) [5][×][7][x²]
   (b) [5][+][7][x²]
   (c) [1][0][0][÷][5][x²]

14 (a) 12.06 (b) 625   (c) 60.25
   (d) 3721   (e) 171.396 (f) 3.56
   (g) 7.9   (h) 1.5   (i) 36
   (j) 7.44   (k) 30   (l) 9

15

1:1	0	2:5		3:1
2		4:3	2	8
		3		
5:4	7	6		6:7
9		7:1	9	8

What progress have you made? (p 387)

1 (a) 20   (b) 1   (c) 14   (d) 50
  (e) 32   (f) 6   (g) 10   (h) 8
  (i) 10   (j) 2   (k) 14   (l) 32
  (m) 8   (n) 16

2 (a) 5.33 (b) 3.9   (c) 2.1
  (d) 329   (e) 14.4 (f) ⁻2.5
  (g) 40   (h) 90

Practice booklet

Sections A and B (p 119)

1 (a) 6   (b) 11   (c) 17   (d) 6
  (e) 11   (f) 7   (g) 13   (h) 3

2 (a) 11   (b) 4   (c) 6   (d) 6
  (e) 12   (f) 6

3 (a) Right   (b) Wrong, 18
  (c) Wrong, 25   (d) Right
  (e) Right   (f) Wrong, 7

4 (a) 18   (b) 15   (c) 4
  (d) 3   (e) 15   (f) 2
  (g) 4   (h) 5   (i) 8

5 (a) 15   (b) 19   (c) 11   (d) 2
  (e) 11   (f) 3   (g) 8   (h) 19
  (i) 2

6 B 2 × (5 + 3)   C 8 × (5 − 3)
  E (1 + 3) × 4

7 A, B, D and E

Section C (p 120)

1 (a) $12 + \frac{6}{2}$ (b) $\frac{12}{6} + 2$
  (c) $\frac{12 + 6}{2}$ (d) $\frac{12}{6 + 2}$

2 A and G, B and H, C and E, D and F

3 (a) 4   (b) 12   (c) 3   (d) 8
  (e) 4   (f) 3   (g) 5   (h) 4

Sections D and E (p 120)

1 (a) 14.84   (b) 20.048 (c) 6.22
  (d) 0.2115 (e) 17.885 (f) 26.5

2 (a) B   (b) D   (c) C   (d) A

3 (a) 20   (b) 3.2   (c) 14.3
  (d) 3.5 (e) 5.25 (f) 5.1
  (g) 8.46 (h) 1.4   (i) 3.62
  (j) 0.04 (k) 0.59 (l) 3.48

4 (a) (i) 2.27 kg (ii) 63.50 kg
      (iii) 2.95 kg
  (b) 1.2 pounds, 0.68 kg, 9 pounds,
      4.09 kg, 4.5 kg

Sections F and G (p 121)

1 (a) 196   (b) 841   (c) 16 384
  (d) 640 000   (e) 20.25   (f) 174.24
  (g) 0.25   (h) 10.1761

2 225, 256, 289 (15^2, 16^2, 17^2)

3 (a) 1521 (39^2)   (b) 1369 (37^2)

4 (a) 21   (b) 51   (c) 1.7   (d) 0.47

5 (a) 63 (b) 42 (c) 3.5 (d) 13.1
6 (a) 18 (b) 4.5 (c) 47
7 Yes, 26 stones

Sections H and I (p 122)

1 (a) 25 (b) 3   (c) 36 (d) 12
  (e) 4   (f) 10 (g) 16 (h) 4

2 (a) 0.74   (b) 6561 (c) 14.06
  (d) 0.092 (e) ⁻7.3 (f) 6.9
  (g) 11.6   (h) 500   (i) 2

54 Chocolate (p 388)   7C/10

This is a problem-solving activity in which pupils can use their
knowledge and understanding of fractions, decimals and/or percentages.

Essential	Optional
6 bars or blocks of something which can be divided up and shared out equally	Bars of chocolate (of a kind not already subdivided into portions)

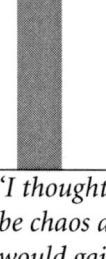

Practice booklet pages 123 to 125

T

◊ The activities and problems are all variations of this basic idea:
  - A number of tables are set out with some chocolate bars.
  - A group of pupils are asked, one by one, to choose a table to sit at.
  - When everyone has sat at a table, the bars on each table are shared equally between those at that table.

The problem for the pupils is to decide which table to sit at in the hope of getting the most chocolate at the shareout.

Getting started

◊ It is best to start with a fairly simple situation. For example, distribute 6 chocolate bars on three tables as shown.

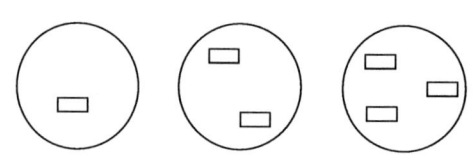

(or start with 2 tables and fewer bars)

'I thought this would be chaos and pupils would gain little from it. In fact, lots of useful discussion was generated! Had I remembered to buy the chocolate it would have been even better. (We used sheets of paper to represent chocolate.)'

Choose a group of pupils to take part, say eight, and explain the problem. Ask them one by one to choose their table (they cannot change their minds later).

◊ As pupils choose where to sit, involve the whole class and ask questions such as:
  - Where would you sit? Why?
  - How much chocolate would each person get at this table if no one else sits here?
  - Is it best to be the first to choose, the last, or doesn't it matter?

'I split the class into three teams. Each team in turn sent one person to sit at the tables. The aim was for one team to get the most chocolate.'

◊ Once the last pupil has chosen, ask pupils to decide who gets the most chocolate and to justify their answer. Explanations could involve
  - comparing fractions, decimals or percentages

> '[The activity] led to whole class discussion on equivalent fractions and converting to percentages – both of which had been covered previously – good enjoyable consolidation.'

> 'In the second lesson, 8 pupils simultaneously sat where they thought they'd get most chocolate. I gave them a chance to move. The rest of the class was the audience. This generated considerable discussion.'

- comparing ratios
- imagining the bars made of an appropriate number of squares and comparing the number of squares each pupil gets
- imagining the bars are a particular weight and comparing the weight of chocolate each pupil gets

Pupils can choose their own approach or you can ask them to concentrate on one aspect, say fractions or decimals.

Variations

◊ Obviously the number of tables and bars can be varied.

You could also tell pupils that you will decide beforehand how many pupils will sit down but they will not know this until you stop them and they then share out the chocolate.

Follow-up work

◊ Pupils could work in small groups on particular problems. Examples are
- For the last person to sit down from a group of eight, what are all the possibilities? Are some of these possibilities more likely than others?
- Which is the best table for the first person to sit at?
- Which final seating arrangements give everyone more than half a bar of chocolate?
- Which seating arrangements are the 'fairest'?

◊ The practice booklet provides some follow-up questions but you may need to select. For example, questions involving fractions will not be appropriate for pupils who have used decimals throughout.

Practice booklet (p 123)

Explanations are given here using fractions but other explanations are possible (for example, using percentages).

1 $\frac{1}{2}$

2 25%

3 Table P: $\frac{1}{2}$ Table Q: $\frac{3}{4}$

4 (a) Q   (b) P

5 Join table A to get the most chocolate. Joining table A would give you $\frac{2}{3}$ of a bar but joining table B would only give you $\frac{3}{5}$ of a bar.

6 (a) Join table P to get the most chocolate. Joining table P would give you $\frac{3}{4}$ of a bar but joining table Q would only give you $\frac{2}{3}$ of a bar.

(b) Join table Q to get the most chocolate. Joining table Q would give you $\frac{1}{2}$ of a bar but joining table P would only give you $\frac{1}{3}$ of a bar.

(c) Join table Q to get the most chocolate. Joining table Q would give you $\frac{1}{3}$ of a bar but joining table P would only give you $\frac{3}{10}$ of a bar.

(d) Join table P to get the most chocolate. Joining table P would give you $\frac{9}{10}$ of a bar but joining table Q would only give you $\frac{4}{5}$ of a bar.

7 6 people

8 (a) 9 people   (b) 8 people

9 (a) 12 bars   (b) 3 bars

10 3 at table P and 6 at table Q

11 12 at table P and 8 at table Q

12 2 at table P and 3 at table Q giving 5 people in total. They will get 2 bars each.

13 On the left, each person gets $\frac{3}{4}$ of a bar. On the right, if 5 people sat down they would each get more than $\frac{3}{4}$ of a bar; if 6 people sat they would get less than $\frac{3}{4}$ of a bar, so it can't be done.

Review 6 (p 388)

Essential
Tracing paper, graph paper, calculator

1 (a) (i) Reflection in the vertical axis
  (ii) Translation 2 units right, 6 units down
  (iii) Rotation of 180° about (1, ⁻3)
 (b) (i) Reflection in the horizontal axis
  (ii) Rotation of 180° about (0, 0)
  (iii) Translation 2 units left, 6 units up

2 (a) 3, 6, 9, 15, 24 (b) 1, 3, 6
 (c) 3, 17 (d) 1, 9

3 (a) £4.80 (b) £4.20
 (c) Group A had the highest mean so it could be said to have done better.

4 (a) ⁻11 (b) ⁻5 (c) ⁻5 (d) 11
 (e) 11 (f) ⁻11 (g) 5 (h) ⁻5

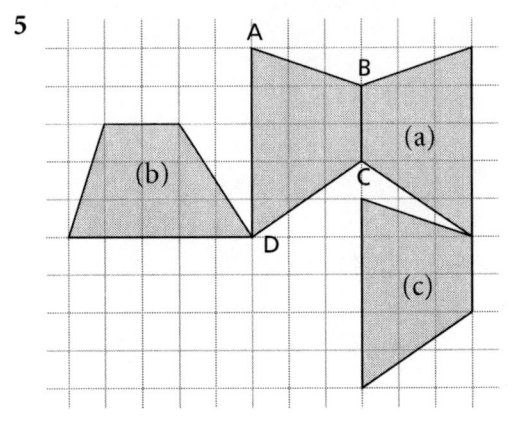

5 [diagram showing shapes labelled A, B, C, D with regions (a), (b), (c)]

6 (a) The numbers in the second row of the table are 5, 7, 9, 11, 13, 15, 17
 (b) Graph of $d = 2t + 5$ (a straight line from (0, 5) to (6, 17))
 (c) 12 centimetres
 (d) $1\frac{1}{2}$ minutes
 (e) $d = 2t + 5$

7 (a) 20 (b) 4

8 (a) The pupil's graphs of $y = x + 3$, $y = 2x$, $y = 8 - x$, $y = 2x + 1$
 (b) $y = 2x$ and $y = 2x + 1$

9 (a) 19 (b) 6 (c) 1 (d) 18

10 (a) Wednesday (b) 12 degrees
 (c) 0.1°C

11 (a) 5 (b) 25
 (c) 2482 (d) 99.28

12 (a) 106 (b) 0.6 (c) 2.56

13 (a) (i) 18.5 (ii) 12 (iii) 13
 (b) The mode or median best represents the age of the group.
  Over 80% of the group is younger than the mean of 18.5 years so the mean is not the most representative.

Mixed questions 6 (Practice booklet p 126)

1 4.8

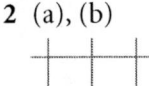

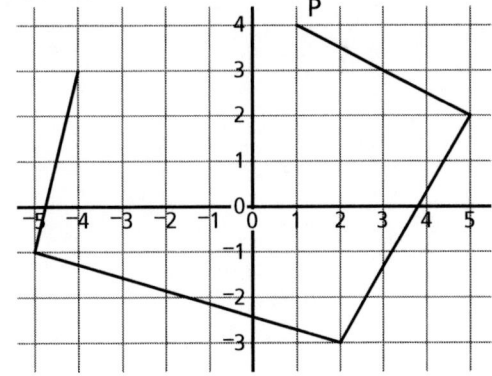

2 (a), (b) [graph showing point P and a quadrilateral plotted on coordinate grid]

 (b) (⁻4, 3)
 (c) 5 units right, 1 unit up

3 (a) ⁻3 (b) ⁻3 (c) 7 (d) 3
 (e) ⁻13 (f) 15 (g) ⁻3 (h) 7

4 (a) $24 \div 4 + 2$ (b) $24 - 4 \times 2$
 (c) $24 - 4 \div 2$ or $24 - 4 + 2$
 (d) $24 + 4 \times 2$

5 (a) False   (b) True
   (c) False   (d) False
   (e) False   (f) False
   (g) False   (h) True

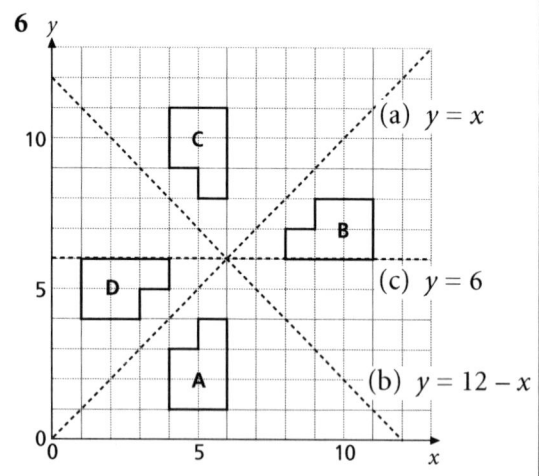

6
(a) $y = x$
(b) $y = 12 - x$
(c) $y = 6$
(d) Rotation 90° anticlockwise about point (6, 6)

7 (a) 6   (b) 24.3   (c) 6.5
   (d) 7.5   (e) 8.5   (f) 27

8 (a) 3   (b) 30   (c) 1   (d) 77

9 (a) 4   (b) 6
   (c) 17 days   (d) 13 days
   (e) 10 days   (f) 10.8 days

*10

¹4	4	²1	■	³1
9	■	⁴1	9	6
■		4		■
⁵3	2	4	■	⁶8
6	■	⁷9	6	1